Richard Kearton

British Birds' Nests

How, where and when to Find and Identify them

Richard Kearton

British Birds' Nests
How, where and when to Find and Identify them

ISBN/EAN: 9783337416898

Printed in Europe, USA, Canada, Australia, Japan

Cover: Foto ©berggeist007 / pixelio.de

More available books at **www.hansebooks.com**

BRITISH BIRDS' NESTS

HOW, WHERE, AND WHEN TO FIND AND IDENTIFY THEM

BY

R. KEARTON, F.Z.S.

AUTHOR OF "BIRDS' NESTS, EGGS, AND EGG COLLECTING"

WITH AN INTRODUCTION BY

R. BOWDLER SHARPE. LL.D.

ILLUSTRATED FROM PHOTOGRAPHS BY

C. KEARTON

OF NESTS, EGGS, YOUNG, ETC., IN THEIR NATURAL SITUATIONS AND SURROUNDINGS

CASSELL AND COMPANY, LIMITED

LONDON, PARIS, NEW YORK & MELBOURNE

1898

First Edition, *September* 1895.
Reprinted March 1896, 1898.

INTRODUCTION.

THE illustrations of BRITISH BIRDS' NESTS which have been submitted to me by Mr. Kearton deserve more than a passing acknowledgment. This book certainly marks an era in natural history, just as Gould's "Birds of Great Britain" and Booth's "Rough Notes" did in the past. The method of illustrating works on natural history has undergone as much development, as the illustration of the animals themselves has done in our public museums. The works of the early part of the century were embellished with the faithful woodcuts of Bewick, or with coloured pictures of more or less merit. These were succeeded by the writings of Macgillivray, Yarrell, and Hewitson, the former containing the best life histories of our British birds, equalled only by those of the great German naturalist Naumann. The great merit of Macgillivray's work lies in the fact that it is almost entirely original, both as regards the descriptions of structure and plumage, as well as of the habits of the birds. Yarrell was also a great naturalist, and his work was illustrated by the neatest of little woodcuts, which survive unto this day; but a candid critic must admit that the attitudes of many of the birds are strained and unnatural, and must have been taken from stuffed specimens. Hewitson's illustrations of the eggs of British birds are still, in my humble opinion, the most beautiful of any which have appeared in this country, though the most perfect representations of birds' eggs yet published are those of Captain Bendire, issued by the Smithsonian Institution.

The influence of faithful illustration in works on natural history soon began to make itself felt in museums, and a higher aim in the efforts of taxidermists became apparent; but the man who most fully realised this necessity, in England at least, was John Hancock, and, after him, the late E. T. Booth, of Brighton, whose museum in the Dyke Road is one of the attractions of that flourishing watering-place. I cannot vouch for the truth of the statement, but I have more than once heard it said that the formation of this collection cost the late owner sixty thousand pounds from first to last. Travelling all over the United Kingdom, and collecting diligently, Mr. Booth managed to get together a very complete collection of British birds, which, however, he did not mount in the usual way of museums, on rows of stands, but placed his treasures in cases, in which the birds were represented with their natural surroundings. Thus we see the Waders feeding on the shore, with a view of the sea beyond; and Stone-Chats sit on their native gorse, instead of perching on a wooden stand, with a very evident stiffness due to the taxidermist's wire. The pictures of bird-life in the Dyke Road Museum are faithfully reproduced by Mr. Neale in the illustrations to the "Rough Notes," which were published by Mr. Booth.

The crowds of people, increasing year by year, who visit the Natural History branch of the British Museum at South Kensington, testify to the popularity of the bird-groups in the national collection. These faithfully represent the natural history of the species, for the actual birds are there, with their nest and eggs or young ones, exactly as they were on the day of their capture; every leaf, every flower, being exactly reproduced. It must, however, not

be forgotten that much of the interest felt in the
nesting habits and the plumage of the young
of our British birds is due to the late John
Gould, in whose magnificent work on the " Birds
of Great Britain" both these features were made
conspicuous. His collection has passed into the
British Museum, and the series of nestling birds is
quite remarkable.

The purchase of such works as those of Gould
and Booth is beyond the compass of most of us, and
this is an age when everyone expects knowledge to be
dispensed at a cheap rate, and to be brought within
the reach of people of moderate means. Even Mr.
Seebohm's work on British birds, the only one
which can fairly be said to take rank beside those
of Naumann and Macgillivray, is expensive, and
unattainable by students of natural history, who are
daily increasing in number. Mr. Kearton, therefore,
steps in at the right moment with this book on
BRITISH BIRDS' NESTS, and it will be some time
before he finds a rival; for the photographs with
which he and his brother have embellished the book
are not only beautiful as photographs, but show us
the nests and eggs of our birds *in situ*. I will not
detract from the interest of the work by quoting
from it; but the way in which these young naturalists have overcome the very serious difficulties
presented by the task they undertook, proves that,
in addition to the native British pluck, the true love
of natural history is necessary to accomplish such a
result as they have achieved. It is everything to
show Nature as she really is, and here photography,
the handmaid of science in the field, comes in.
Artists will undoubtedly admire the illustrations, but
the naturalist will love them still more, because they
show him the nests of the birds as the authors
A*

discovered them; no imaginary details, as is so often
the case in illustrations of bird-life, but the actual
nest itself, so that the perusal of the book is a birds'-
nesting expedition. The descriptions of the birds
and of the material of the nests are both adequate
and instructive, and the latter will be most useful.

I am certain that we are not going to hear the
last of Mr. Kearton and his brother with the present
volume; and, if I may suggest, naturalists would
especially like to hear more of their birds'-nesting
experiences in detail. There is so much still to
be learnt about the habits of birds, even of our
commoner kinds; and in a long course of editing
and writing of books on birds, I find the greatest
difficulty in gathering any new facts about their
habits. Even in my suburban garden I have
learnt many interesting facts; and anyone who has
travelled, like Mr. Kearton, in search of subjects for
the pen and camera, must have much to tell. That
he could relate his experiences excellently, too, is
evident from his preface to the present work, and
I can only hope that he will speedily do so. The
difficulties which he and his brother have sur-
mounted in procuring their series of beautiful photographs of British birds' nests, prove that there is
nothing which they would not dare; and I can assure
them that there is nothing for them to fear from
the British public, who will undoubtedly be glad to
hear some more of their stories of bird-life in the
field.

R. BOWDLER SHARPE.

British Museum (Nat. Hist.),
South Kensington,
Sept. 7, 1895.

PREFACE.

THE primary object of this work is to supply such information as will help the student of ornithology to find and identify the nests and eggs of birds breeding within the British Isles. In its preparation I have endeavoured always to keep my own early wants in view, and worked accordingly.

The arrangement of the book (a sort of combination of Montagu and Newman) is, I am inclined to think, the best for a popular work of its kind, and enables a ready reference to any bird or its breeding habits and economy. The brief descriptions of the parent birds—which are as far as possible in their breeding plumage, and have been made as concise and practical as circumstances would admit— will be found of considerable assistance in identifying specimens, which is unquestionably the most important point connected with the study of the subject.

I recognised this fact when I was a lad of nine years of age, and in endeavouring to carry it to its logical conclusion brought considerable trouble upon myself. During one of my solitary excursions along the side of a noisy Yorkshire beck I came across a bird's nest, differing so widely from all my previous "takes" and discoveries that I determined to find out the species to which it belonged, and accord-

ingly hid myself and began to watch for the return
of its builder. Night fell without bringing any
success, so I curled myself up beneath an overhanging crag, in order to wait until the bird came
back in the morning to lay another egg, entirely
oblivious of the fact that I should be missed
at home. My slumbers were broken early by a
great outcry in the little gill. The whole of the
able-bodied population of our mountain village had
turned out to help find me, and I have good reason
for believing that many mournful prophecies as to
my fate were indulged in by those who knew something of my queer habits. I suffered that night somewhat severely in the interests of science; nevertheless
my ardour was not quenched, and I returned at
the earliest opportunity to the strange nest by
the beck-side, when a sight of its owner as she
hurriedly left her eggs made it plain that the domed
house of moss belonged to a Dipper. I have on
many occasions since that gratified my desire to
spend a night with the birds, both on land and sea,
in winter and in summer, and have learned on
each occasion something it was both pleasant and
profitable to know.

I have been a bird lover and collector for upwards
of twenty years, and much of this work has been
written from my own specimens and note-books,
giving, of course, in the case of the birds themselves,
due care to the parts that fade and alter after
death. I have also freely consulted and unreservedly
acknowledge my indebtedness to such authorities
as Yarrell (revised and enlarged by Messrs. Newton

and Saunders), Seebohm, Dresser and Sharpe, Dixon, Montagu, Morris, Newman (revised and rewritten by Miller Christy), Stevenson, Swaysland, Booth, Bechstein, Atkinson, and others. Neither must I forget to mention the *Field*, from the natural history columns of which I have compiled a whole note-book of useful and extraordinary facts, many of which find a place in the following pages.

I have also to express our thanks, for valuable assistance readily rendered, to the Duke of Argyll, the Marquis of Lorne, Lord Walsingham, the Conservators of Epping Forest, Mr. J. Anderson of Oban, Mr. H. A. Paynter of Alnwick, Mr. James Sinclair of Kirkwall, Mr. Robert Baldry of Norwich, Mr. R. J. Ussher of Cappagh, Mr. C. H. Bisshopp of Oban, and the numerous factors, farmers, keepers, and boatmen who have assisted us most willingly in obtaining the necessary materials for the production of the work.

The great feature of the book lies in the unique character of its pictures. In this respect we can claim that it is the first practical attempt to illustrate a manual on the subject from photographs taken *in situ*. A glance through its pages will at once establish the valuable nature of this new departure. Of course, it must not be supposed that all the nests and eggs are to be seen exactly as represented in the illustrations; for many of them had to be partially exposed before a photograph could possibly be taken, and in some instances actually removed from holes, as in the case of the Wheatear, Starling, and Swallow.

The experienced ornithologist will no doubt miss a few nests he might reasonably expect to find in this work; but on the other hand it is certain he will discover some of infinitely greater value, which he did not expect to fall in with. The omission of such birds as the Golden Plover, Dabchick, Yellow Wagtail, Corncrake, and a few others, is due in some cases to accident, and in others to our inability to find them, although we searched long and diligently. In some instances it happened that directly we succeeded in finding and photographing a bird's nest which had entailed considerable trouble in the search, we immediately came across others by accident. This was notably the case with the Ringed Plover, Kingfisher, and Dunlin. In several cases good pictures have been left out because we were unable, from lack of trustworthy evidence, to identify them with absolute certainty.

No one who has yet to try this particular branch of photography, can possibly appreciate its troubles and disappointments. As an instance of the latter, my brother on one occasion travelled upwards of five hundred miles by rail, and dragged his camera at least twelve miles up and down a mountain side, in order to take a view of one bird's nest, and was defeated by the oncoming of a thick mist at the very moment he was fixing up his apparatus. The Golden Eagle's eyrie was photographed during the temporary lifting of a Highland mist, and considering the situation, and the unsatisfactory state of the light, has turned out very successfully.

The picture of the Solan Goose was obtained

about four o'clock in the morning on Ailsa
Craig, and so early in the season that the birds
had not settled down seriously to the business of
incubation; and is of especial value and interest
to us on account of the adventures we encountered
on that "beetling crag." In getting down to the
edge of the cliff, my brother placed too much
dependence upon the stability of a large slab of
rock, which treacherously commenced to slither
down the terribly steep hill side at a great pace
directly it received his additional weight. He
narrowly managed to save himself and the camera,
with which he was encumbered at the time, from
being shot over the lip of the precipice, and sustain-
ing a fall of several hundred feet, into the sea below.
We took five photographs of the Gannet sitting
on her nest, each at closer range, and although
she was ill at ease while all this was going on, by
working deftly we established ourselves somewhat
in her confidence, and got close enough to obtain
the picture forming the frontispiece to this work.
When everything was ready, as if by the malicious
intervention of some unkind fate, the screw affixing
the camera to the tripod suddenly dropped out, and
the apparatus toppled over seawards. It was well on
its way to what the Americans describe as "ever-
lasting smash," when my brother, by a dexterous
catch, stopped it from striking a piece of rock, off
which it would have rebounded and finally dis-
appeared over the cliff. By the aid of some strong
feather shafts (the only materials available), we
managed to fix up, after a fashion, our apparatus

again, and whilst the artist held the camera on to the tripod, and the author, from a more secure footing, held the artist by the coat-tails on to the Craig, the picture was obtained, which I venture to think amply rewards us for our trouble.

The members of an eminent Northern natural history society visited the Craig about a fortnight afterwards for the same purpose, and, from the insurmountable character of the difficulties which presented themselves, had to return empty-handed. Nobody who has essayed the same task will much blame them.

The Kestrel's nest was situated on the stump of a tree growing at right angles from a cleft in a Highland "scar," and, as may be judged, was in an exceedingly awkward situation to photograph. However, by the kindly assistance of a gamekeeper and a strong rope, my brother and his camera were lowered on to the stump, sitting astride of which he lashed his apparatus thereto and exposed four plates, every one of which turned out a failure on account of the close range, snowy whiteness of the uneasy little birds, and the blackness of the background. We were obliged to do the whole thing over again, and after a series of experiments managed to secure the photograph from which our illustration has been reproduced.

In 1894 we were unable to land upon the Meg-stone Rock at the Farne Islands, for the purpose of securing a picture of the Cormorants' nests, on account of the dangerous swell running at the time. This year we went considerably out of our way in order to remedy the defect in our series, and after inducing the boatman and his crew to land us, at great risk to the craft and all our lives, were disappointed to discover that a recent north-easterly gale had swept every nest and egg off the rock. The birds were busy building new nests, and our illustration represents the foundations of some of them.

We sought hard on the shores of several Highland lochs for the nest of a Sandpiper, from four o'clock in the morning till breakfast time, three or four days running, but in vain; and when we had almost given up the task as hopeless, a curious thing happened. I was lying at full length one evening upon a great fang-like promontory of rock that jutted out into a sea loch, testing the capabilities of a small rifle, when I startled a bird off her nest, containing four eggs, in a tussock of rushes close by me. At this time we had exposed all our stock of plates except two, and the reader will be able to judge of our chances of renewing supplies, when I tell him that we were staying at a place so remote and isolated that the kind Highland body who put us up (at considerable inconvenience to herself and her family) was for the space of three or four days quite unable to procure us fish, flesh, or fowl to eat. A strong breeze was blowing at the time from seaward, and after weighting the camera as heavily as we dared with a lump of rock to prevent vibration, and sheltering the waving rushes with our jackets, one of the remaining precious plates was exposed.

Fearful that we had not succeeded, my brother
returned to our quarters a mile away and developed
the negative—alas! only to find that the vibration
of the apparatus had utterly ruined it. We were
now reduced to our last chance. The breeze had
increased if anything, and the prospect of obtaining
a decent picture certainly looked gloomy. We
rammed the legs of the tripod as far into the
hard-baked handbreadth of ground as possible,
weighted each with a heavy leaning stone, hung
a huge iron otter trap beneath the body of the
camera, and placed as big a stone as the construction of the machine would bear on the top; then
waited, jackets in hand and shoulder to shoulder,
for a temporary lull in the wind. Luckily the
result rewarded our trouble.

The Kingfisher's nesting-hole was in such a
position that my brother was obliged to photograph it standing waist-deep in the river Mole, the
swirling waters of which reached very nearly up
to the body of the camera. The nests of the Ring
Dove, Sparrow Hawk, and several other high builders
were obtained by climbing adjoining trees and
lashing the apparatus thereto, building scaffolds
and other contrivances, the difficulties of which a
practical photographer alone can fully appreciate.

Some of the pictures were obtained with one of
Messrs. Dallmeyer's tele-photo lenses, an admirable
contrivance for inaccessible objects and timid
animals, which it would be impossible to approach
near enough with an ordinary apparatus. Despite
its difficulties and dangers, photographing natural

history objects in their own haunts and elements
is an extremely pleasant pastime, and no one
pursuing it should be without one of these instru-
ments, which we had made adjustable to a half-
plate camera. A good pair of field-glasses are
also very useful for watching shy birds on or off
their nests, and I have made some really astonishing discoveries by the aid of mine at times.

The circumstances governing the distances at
which our photographs were taken made it quite
impossible, as may be easily understood, for us to
adjust the relative sizes of the eggs of different
species. However, the average measurement of
each bird's egg in inches and the decimal parts
thereof will easily prevent any misconception on
this point.

As a great deal of valuable time is often wasted
in finding out how to get to good sea-bird haunts,
I propose to give particulars respecting some of
the best we have visited.

The Farne Islands are now under the control
of an Association, having for its *raison d'être* the
protection of the birds breeding on these far-famed
rocks. However, passes may be purchased from
Mr. Hughes, landlord of the Ship Inn, North
Sunderland (nearest railway station — Chathill).
Fisherman Richard Allen has a capital craft for
taking visitors off, and is a capable and experienced
boatman—a distinct acquisition to the naturalist or
photographer, where tides and currents run as fiercely
as mill-races. Mr. Kendal, of the Canty Bay
Hotel, some four miles from North Berwick, rents

the Bass Rock; but before paying a contemplated visit to that ancient haunt of the Solan Goose, it will be well to make prior arrangements, in order to save time and prevent disappointment.

Ailsa Craig is rented by two brothers named Girvan, one of whom lives on the rock and the other at Girvan, on the Ayrshire coast. The one upon the mainland will ferry the visitor across the nine miles or so of ocean, and the brother on the Craig will act as guide over the awesome rock-stack.

Oban forms a very good base of operations for doing the Hebrides (per Mac Brayne's splendid service of boats) and the mountains round about. John Cowan, boatman, of Corran Esplanade in Oban, is a very intelligent fellow from an ornithological point of view, and owns a good sailing boat wherewith the small islands in the adjoining lochs may be profitably worked. The Shiant Islands may be visited by a hired boat from Tarbert, Harris; however, we experienced considerable difficulty at this place, by reason of the men being actively engaged in fishing just when we required their services.

In conclusion, I need hardly say that we shall be extremely grateful to any gentleman kind enough to give us an opportunity of taking the photograph of any bird's nest which does not already appear in this work.

Boreham Wood,
Elstree, Herts, 1895.

BRITISH BIRDS' NESTS.

ACCENTOR. *See* HEDGE SPARROW.

BITTERN.

Has now quite ceased to breed in the British Isles, and is only a rare visitor.

BLACKBIRD.

Description of Parent Birds.—Length about ten inches. Bill of medium length, nearly straight, and yellow. Eyelids yellow. Plumage uniform deep black. Legs and toes brownish black, claws black.

The female is of a dark rusty brown colour, bill and feet dusky brown.

Situation and Locality.—Bushes, hedges, ledges of rock, holes in and on projecting " troughs " of dry walls, on banks, in evergreens, against the trunks of trees, and occasionally quite on the ground. I have upon more than one occasion found a Blackbird's nest built upon an old Thrush's, and *vice versa.* Common nearly all over the United Kingdom.

Materials.—Small twigs, roots, dry grass, moss, intermixed with clay or mud. Sometimes bits of wool, leaves, fern fronds, and even paper, lined internally with fine dry grass.

B

Eggs. — Four to five, sometimes six. Some authorities say as many as seven and eight; but although I have been birds'-nesting now for five-and-twenty years, I never met with either number. Of a dull bluish-green, spotted and blotched, and rarely streaked with reddish-brown and grey. They vary considerably, both in regard to ground colour, shape, size, and markings. Some varieties are covered with small spots, others with such large ones that they very closely resemble the eggs of the Ring Ouzel, whilst a third variety is almost spotless. Size about 1·18 by ·85 in.

Time. — March, April, May, June, July, and even as late as August.

Remarks. — Resident. Notes: call, *tsissrr*, *tuck*, *tuck;* alarm, a loud, ringing *spink*, or *chink*, *chink*, *chink.* Song powerful, and generally delivered at the beginning or end of the day. Local and other names : Merle, Black Ouzel, Amzel Ouzel. A close sitter.

BLACKCAP.

Description of Parent Birds. — Length about six inches ; bill of medium length, straight, and dark horn colour. Irides brown. All the upper part of the head black ; nape ash grey ; back and wing-coverts ash-grey, tinged with brown ; wing and tail-quills brown, bordered with grey ; cheeks, chin, throat, and breast light grey; belly and under-parts white ; legs and toes lead colour ; claws brown.

The female is larger than the male ; the top of her head is dull rust colour, and her plumage generally more tinged with brown.

Situation and Locality. — In brambles, briars, thick hedges, nettles, and gooseberry bushes, in

gardens, orchards, thickets, shrubberies, and other suitable places, at varying heights of from two to ten or twelve feet from the ground. In all parts of England and Wales, and more sparingly in Scotland and Ireland. It is a very widely distributed bird.

Materials.—Fibrous roots, straws, and dead grass, with an inner lining of hair. It is a flimsy structure, sometimes strengthened by wool or spiders' webs.

Eggs.—Five to six, very variable, and often difficult to identify, as they closely resemble those of some of the other Warblers. The commonest type is that of a greyish-white underground, suffused with buffish-brown, and spotted, blotched and marbled with dark brown. Sometimes they are found pale brick-red, marked with a darker tinge of the same colour, and reddish-brown; also faint blue, marked with grey and yellowish-brown. Size about ·78 by ·58 in.

Time.—May and June.

Remarks.—Migratory, arriving in April and leaving in September, odd specimens remaining till November or December. Notes: alarm, *tack, tack,* or *tee, tee.* Song of great power and freedom. Local and other names: Hay Jack, Hay Chat, Mock Nightingale, Nettle-creeper, Nettle-monger, Blackcap Warbler. Sits very closely.

BULLFINCH.

Description of Parent Birds.—Length about six inches. Bill short, broad, and thick at the base, and black. Irides dark brown. Round the base of the beak, and all the upper part of the head, black. Nape, back, and lesser wing-coverts grey;

BULLFINCH.

greater coverts black, tipped with greyish-white,
which forms a bar across the wing-quills, dusky.
Rump white, upper tail-coverts and tail-quills black.
Cheeks, breast, and belly tile-red. Vent and under
tail-coverts white. Legs and toes flesh colour;
claws brown.

The female has the black on the head, wings,
and tail not so intense; her nape and back are
greyish-brown, and breast, belly, and under-parts
dirty brown.

Situation and Locality.—In the lower branches
of trees, the tops of high bushes, in thick, quick-
set hedges and thickets in suitable localities throughout the British Isles; rarer towards the extreme
north of Scotland and in Ireland. The one represented in our illustration was situated in a thick
Surrey wood.

Materials.—Small twigs and fibrous roots, interlaced so as to form, as a rule, a broad and flat
platform, in the centre of which is the cup-shaped
recess lined with fine fibrous roots and sometimes
a little wool, hair, or a few feathers.

Eggs.—Four to six. Pale greenish-blue, spotted,
speckled, and sometimes streaked with dark purplish-
brown, and with underlying blotches of brownish-
pink. The markings generally form a zone round
the large end of the egg. Size about ·77 by
·57 in.

Time.—April, May, June, and July.

Remarks.—Resident. Notes: call, soft, plaintive,
and frequently-uttered; song, feeble and low. Male
ceases to sing as soon as eggs have been laid.
Local and other names: Beechfinch, Horsefinch,
Pink, Twink, Olph, Nope, Red Hoop, Alp, Hoop.
Sits very closely indeed.

BUNTING, CIRL.

Description of Parent Birds.—Length about six inches. Bill short, conical, and bluish-grey. Irides hazel. A streak of light yellow runs from the base of the upper mandible, over the eye, behind the ear-coverts, and thence forward under the black on the throat; and another from the gape under the eye; crown, behind the eye and under the lower yellow streak, olive-brown streaked with black ; the head, nape, and sides of the neck dark olive ; back rich chestnut-brown, the margins of the feathers being tinged with olive ; wings dusky black edged with chestnut-brown and olive green ; upper tail-coverts olive, tinged with yellow and marked with dusky grey streaks; tail-quills brown, outside feathers edged with white; chin and throat black ; upper part of breast dull olive, crossed below by a chestnut band; belly and under tail-coverts dusky yellow ; sides dull olive streaked with dark brown.

The female is slightly smaller; she lacks the bright yellow stripes on the sides of the head and throat; the black upon the chin is replaced by yellowish brown ; crown dull olive streaked with black ; back and upper parts reddish-brown streaked with black ; under-parts dirty yellow, also streaked with black.

Situation and Locality. — In brambles, furze bushes, and sometimes quite on the ground, in similar situations to the Yellow Bunting. On commons and cultivated lands well studded with trees and hedgerows in the South and West of England.

Materials.—Dry grass, roots, moss, and leaves, with generally an inner lining of hair.

Eggs.—Four to five. Dull bluish or cinerous white, spotted, blotched, streaked, and veined irregularly with very dark brown, and underlying markings of grey. Size about ·86 by ·65 in. *Time.*—May, June, and July. *Remarks.*—Resident. Notes: call, *zi-zi-za-zirr*; song, *zis-zis-zis-gör-gör-gör*, according to Bechstein, whilst another authority describes it as *tütt, tütt, tütt, tütt, tütt, tütt*. Local and other names: Black-throated Yellow-hammer, French Yellow-hammer. Sits close.

BUNTING, COMMON. *See* BUNTING, CORN.

BUNTING, CORN. *Also* COMMON BUNTING.

Description of Parent Birds.—Length a little over seven inches. Bill short, conical, strong, and pale yellow-brown, with a stripe of dark brown on the top of the upper mandible. Irides dark hazel. Head, neck, back, rump, and upper tail-coverts light brown, inclining to olive, each feather being streaked in the centre with dark brown. Wings dark brown, the feathers being edged with a lighter tinge of the same colour. Tail slightly forked and dark brown with light edges. Chin, throat, breast, belly, vent, and under tail-coverts dull whitish-brown; the sides of the neck and the breast are marked with triangular spots of dark brown; sides streaked with the same colour. Legs, toes, and claws pale brown. The bird is thick and bulky in appearance and of sluggish habits.

Female similar to male.

Situation and Locality.— On or near the ground, amongst coarse grass, on a bank, among the grass, under a hedge, in a low bush, amongst growing corn, brambles, clover, and peas; in grass-fields, pastures, clover-fields, and similar places locally distributed throughout the United Kingdom.

Materials.—Straw and coarse hay or grass-stems outside, lined with fibrous roots, fine grass, and sometimes horsehair.

Eggs.—Four to six. The ground colour varies from dull purplish-white to pale buff, blotched, spotted, and streaked with light to dark purplish-brown, and underlying markings of grey. They are variable in size, but run larger than those of any other Bunting breeding with us. Size about ·96 by ·71 in.

Time.—May and June. Individual nests may be, however, found as early as the end of April and as late as the beginning of July.

Remarks.—Resident, but numbers swollen during winter months by Continental arrivals. Notes, *chuck* or *chit.* Local and other names: Common Bunting, Bunting Lark, Ebb. Sits close.

BUNTING, REED. *Also* REED SPARROW.

Description of Parent Birds.—Length about six inches. Bill short, conical, and dusky brown on the upper mandible, lighter on the lower. Irides hazel. Head velvety black, bounded by a white collar, which commences near the gape and, descending the sides of the neck as far as the breast, passes round the back thereof. Back and wings rich brownish-black, the feathers being margined broadly with reddish-brown and tawny-grey; wing-quills dusky, narrowly bordered with tawny-red. Rump and upper tail-coverts black, tinged with rusty grey. Tail-quills brownish-black, bordered on the outsides with white, and slightly forked. Chin and throat black, broad in the centre, and pointed on the lower part of the breast. Breast on either side of the black portion white, also belly and under-parts; sides and flanks tinged and streaked with brown. Legs, toes, and claws dusky brown.

The female is smaller, and differs considerably in her plumage. Her head is brown instead of black; the white collar of the male is dusky-brown in her case. Chin, throat, and breast dull white.

Situation and Locality. — Generally near the ground amongst long grass, rushes, nettles, and sedges. I have once or twice met with it in low thorn bushes, amongst grass and weeds growing about the stunted branches which had been cropped by sheep. Our illustration was procured in July amongst the sprouts and long grass growing round the stump of a felled tree on the banks of the Mole, in Surrey. It is generally found close to sluggish streams, ponds, swamps, and bodies of water, though I have frequently met with the nest at considerable distances from water. It is said to have been found in trees at a height of eight or nine feet. It may always be known from that of the Reed Wren by the fact that it is never suspended. It breeds in nearly all suitable localities throughout the British Isles.

Materials.—Dried grass and moss, with a lining of finer grass, hair, and the feathery tops of reeds. I have a nest before me which is composed entirely of wheat-straws, hay, and white horsehairs.

Eggs.—Four to seven, generally five. Purplish-grey or pale olive to pale purple-brown in ground colour, spotted and streaked with rich dark purple-

REED BUNTING.

brown, generally distributed over the egg. It is the smallest of the Buntings' eggs found in this country, and the veins are shorter and thicker than those of the Yellow Bunting. Size about ·77 by ·57 in.

Time.—March, April, May, June, and July.

Remarks.—Resident, and partially migratory. Notes: song, *te, te, tu, te,* diversified by an occasional discordant *ruytsh:* alarm note, a sharp twitter. Local and other names: Reed Sparrow, Passerine Bunting, Black Bonnet, Chink, Water Sparrow, Black-headed Bunting, Mountain Sparrow. Sits closely.

BUNTING, SNOW.

Description of Parent Birds.—Length about seven inches; bill short, conical, and black. Irides hazel. Head and neck white (in some specimens the crown and nape are mottled with black); back velvety black; rump and upper tail-coverts white, some of the feathers being slightly bordered with brownish-white; wings black on the shoulder or point, white through the middle, and black on outer half and tips; tail slightly forked, white on the outside and black in the middle; chin, throat, breast, belly, vent, and under tail-coverts pure white; legs, toes, and claws black.

The female is rather smaller, has the white on the head and neck more mottled with black, and her colours generally are not so pure. Very few specimens of the bird have been secured in this country in its breeding plumage.

Situation and Locality.—In crevices and chinks of rock, or amongst loose stones. The bird is said to breed on the high hills and mountains of the far north of Scotland, in the Orkneys and Shetlands; but very few nests indeed have ever been found.

Materials.—Dead grass and roots, with an inner lining of finer grass, hair, wool, and feathers where procurable. The same nest is said to be used more than once.

Eggs.—Four to eight, more often four to six; ground colour greenish or bluish-white, sometimes greyish-white, pale bluish-grey, or pale greenish-blue, spotted, splashed, and streaked with deep brownish-red, and a few spots and streaks of a darker tint on the top of these; occasionally under-lying markings of pale grey or yellowish-brown. Size about ·86 by ·64 in.

Time.—May, June, and July.

Remarks.—A winter visitor, a few pairs resident. Notes, sweet and tinkling. Local names: Snow Flake, Snow Fleck, Snow Fowl, Tawny Bunting, Greater Brambling, Lesser Mountain Finch, Great Pied Mountain Finch, Brambling (a name belonging to another bird altogether). Said to be a close sitter.

BUNTING, YELLOW. *See* YELLOW HAMMER.

BUZZARD, COMMON.

Description of Parent Birds. — Length about twenty-two inches. Beak short, much curved, strong, and blue-black in colour. Bare skin round the base of the beak yellow. Irides yellow. Crown, nape, back wing-coverts, and upper side of tail-quills clove or ferruginous brown, with large longi-tudinal spots and dashes; the tail being barred with black and ash-colour, and at the end dusky white.

Wing-quills brownish-black. Chin and throat yellowish or dusky white. Breast, belly, and thighs greyish or yellowish-white, streaked and spotted with yellowish-brown. Under tail-coverts white. Legs and toes yellow; claws black.

The female is darker than the male, and often larger. The colour of plumage in both sexes is subject to great variation.

Situation and Locality.—In the forked branches of a tree, sometimes on a horizontal branch at a little distance from the trunk. Also in high, inaccessible maritime cliffs and tall crags in wild secluded districts of England, Wales, Scotland, and Ireland. The bird will often adopt an old crow's nest, and generally returns to the same breeding-place year after year. Our illustration represents a cliff in Mull, in which a Buzzard and Peregrine Falcon were breeding at the time the photograph was taken.

Materials.—Sticks and twigs in liberal quantities, lined with hay, wool, and leaves, sometimes scraps of down.

Eggs.—Two to four, generally three. Sometimes dingy white and unspotted, at others greenish or bluish-white, spotted, blotched, and streaked with red-brown and pale rust-colour. Very variable in regard to size and coloration. Average about 2.16 by 1·72 in.

Time.—April and May.

Remarks.—Resident. Note, a monotonous and plaintive *pe-e-i-o-oo*. Local and other names: Buzzard, Puttock. Not a very close sitter, except when incubation is advanced.

NESTING-PLACE OF THE COMMON BUZZARD AND THE
PEREGRINE FALCON.

BUZZARD, HONEY.

Though never a numerous species in our islands, this bird did at one time breed in several different parts of the country. It is probable that the high price set upon its eggs, and the senseless persecution of the harmless parent birds, have almost, if not quite, banished the Honey Buzzard from its last stronghold—viz. the New Forest.

BUZZARD, ROUGH-LEGGED.

This bird is said to have bred in both England and Scotland, but as the instances are rare, and the information concerning them scant, it can claim but little attention in a work of this character.

BURROW DUCK. *See* SHELDRAKE.

CAPERCAILLIE.

Description of Parent Birds.—Length varies between thirty-three and forty inches. Bill short, much curved, strong, and of a pale horn colour. Irides hazel. Over each eye is a piece of naked red skin. Head, neck, back, rump, and upper tail-coverts brownish-black in ground colour, finely freckled with ash-grey spots. Wings dark chestnut-brown, minutely speckled with dusky spots, except quills, which are dusky. Where the shoulder or point of the wing meets the body, is a little patch of white; the scapulars are also tipped with the same colour. Tail rounded at the tip, the feathers being dusky, spotted sparingly with light grey on the sides; a few shorter ones lying over the principal quills are tipped with white. Chin and throat dull black, the feathers being somewhat elongated. Upper breast dark glossy green, lower breast and all under-parts black, spotted sparingly with white about the thighs and vent. Legs covered with brown hair-like feathers; toes and claws black.

The female measures about twenty-six inches in length, and differs very considerably in the colour of her plumage. The feathers of her head, neck, back, wings (except quills, which are dusky), rump, upper tail-coverts, and tail tawny brown, barred with blackish-brown and tipped with white. Throat tawny red; breast of a lighter tinge, spotted sparingly with white; belly and under-parts generally, barred with pale tawny and black, the feathers being tipped with greyish-white. Legs greyish-brown; toes and claws pale brown.

Situation and Locality.—On the ground, under a bush or bramble, amongst long grass or heather, in Scotch fir, larch, and spruce forests; also, but more sparsely, in oak and birch forests, through the Midlands of Scotland.

Materials.—Dead grass, leaves, or pine needles used as a lining to the hollow, scraped or chosen, in the ground.

Eggs.—Six or eight to twelve or fifteen. Pale reddish-yellow, spotted all over with two shades of darker orange-brown. Size about 2·2 by 1·6 in.

Time.—April, May, and June.

Remarks.—Resident. This bird became extinct in Britain towards the end of the eighteenth century, and was re-introduced from Sweden in 1837, since which time it has thriven and spread in Scotland. Call of male, *peller, peller, peller.* The note of the

c

female is a hoarse *gock, gock, gock.* Local and other names : Wood Grouse, Ceiliog Coed (of the ancient British), Cock of the Woods, Great Grouse, Cock of the Mountain, Capercally, Capercailzie, Capercali. Sits closely.

- - -

CHAFFINCH.

Description of Parent Birds.—Length about six inches. Bill shortish, strong, conical, pointed, and dark blue. Irides hazel. Forehead black; crown, hinder part of head, and a part of the sides of the neck bluish-ash. Back reddish-brown ; rump and upper tail-coverts greenish. Lesser wing-coverts white; greater black tipped with white, thus forming two conspicuous bars across the wings; quills dusky, bordered with greenish-yellow on the outer webs and marked with greyish-white on both webs near the base. Tail-quills black, tinged with grey on the two middle feathers, and the two outer ones on each side marked with white. Chin, cheeks, throat, breast, belly, and under-parts reddish, chest nut-brown, paler on the belly, vent, and under tail-coverts. Legs, toes, and claws dusky.

The female is smaller, and her head, neck, and upper part of the back greyish-brown : the rump and upper tail-coverts are not so bright, and her under-parts are brownish-white, tinged with red upon the breast. The white bars upon her wings are not so conspicuous.

Situation and Locality.—In the forks of small trees, on branches and twigs of whitethorns, fruit trees, in hedges, gorse bushes, and other kinds of trees in orchards, spinneys, on commons, and almost anywhere and everywhere where there are trees

CHAFFINCH.

or shrubs throughout the United Kingdom. Our illustrations are from photographs taken in widely different parts of the country.

Materials.—Moss, wool, lichens, and cobwebs beautifully felted together, and lined with hair, feathers, and down. The nest is cup-shaped, deep, and wonderfully made in every respect. It is compact, neat, well felted or woven together, and securely fastened to the situation chosen. The bird shows a great deal of sagacity in its outside adornment. I have specimens in my possession taken from lichen-covered or grey-barked trees that are smothered with bits of lichen and spiders' nests, and have seen bits of old newspaper used for the purpose. On the other hand, I have nests whereon none of these materials appear, because their surroundings did not call for them to produce any harmonising effect.

Eggs.—Four to six, generally five. Pale greenish-blue, generally suffused with faint reddish-brown and spotted and streaked with dirty reddish-brown of various shades. I have specimens in my possession suffused with purplish-buff all over, but without any markings; and have seen a clutch of pale greenish-blue ones entirely unmarked. Size about ·75 by ·58 in.

Time.—March, April, May, June and July.

Remarks.—Resident, and partially migratory. I have noticed that the cocks, almost without exception, leave the Yorkshire dales in winter. Notes, *spink-spink*, *yack-yack*, *treef-treef*. Its song is a joyous, ringing trill. Local and other names: Bullspink, Scobby, Skelly, Spink, Twink, Pink, Shellapple, Shelly, Shilfer, Wet Bird, Buckfinch, Beechfinch, Whitefinch, Copperfinch, Horsefinch. A very close sitter. I have seen the bird's tail pulled out by a lad attempting to catch the hen on her nest; yet, quite undaunted, she returned, hatched out her five eggs, and reared her young.

CHIFF-CHAFF.

Description of Parent Birds.—Length about four and three-quarter inches. Bill moderately long, nearly straight, and dark brown in colour. Irides brown. Over the eye is a pale yellowish-brown streak, which becomes much lighter behind the ear-coverts. Crown, neck, back, and upper tail-coverts dull olive-green tinged with yellow. Wing-quills dark greyish-brown edged with olive-green; tail-feathers somewhat similar. Chin, throat, breast, and under-parts dull yellowish-white. Legs, toes, and claws blackish-brown.

The female is very similar in all respects.

Situation and Locality.—On or near the ground, in woods, on hedge-banks amongst tall, rank grass, supported by brambles and slender bushes. Our illustration shows a bank in which is a nest exposed so that the eggs may be seen. It breeds pretty generally throughout the south and middle of England, but less frequently in the northern counties and Scotland. It is also met with in Wales and Ireland.

Materials.—Dead grass, withered leaves, and moss, with an inner lining of hair and a liberal quantity of feathers. The nest is oval, or nearly so, domed, and has an entrance hole at the side and near the top.

Eggs.—Five to seven, white, somewhat sparingly spotted with dark purplish-brown. The spots vary in size, intensity, and numbers, but as a rule, they are darker, fewer, and larger than those found upon the eggs of the Willow Wren. Size about ·6 by ·48 in.

Time.—April, May and June.

Remarks.—Migratory, arriving in March and departing in October. Notes, *chiff-chaff*, which has also been represented as *chip-chop*, *chiry-chary*, *choice and cheap*. Alarm note, *whoo-id* or *whoo-it*. Local and other names: Least Willow Wren, Lesser Pettychaps, Choice and Cheap. A close sitter.

CHOUGH.

Description of Parent Birds.—Length about sixteen inches. Bill rather long, slightly curved, and orange-red. Irides hazel. The whole of the plumage is black, glossed with purple; legs and toes red; claws strongly hooked, and black.

The female is not as large as the male, and her bill is shorter.

Situation and Locality.—In clefts and fissures of sea cliffs, holes in old ruins, and in caves. It breeds now very sparingly in Cornwall, Devonshire, Lundy Island, along the Welsh coast, Isle of Man, and more plentifully on the West coast of Scotland, and some parts of Ireland; but everywhere seems to be on the decrease—a distressing fact which it is to be hoped all British ornithologists will assist to stop.

Materials. — Sticks, dead heather stalks, dry grass, wool, and occasionally hair.

Eggs.—Four to five, occasionally six. Dirty or creamy white, sometimes faintly tinged with green or blue, spotted with light brown and ash grey. Markings variable, both in regard to size and distribution. Size about 1·52 by 1·1 in.

CHIFF-CHAFF.

Time.– May.

Remarks.—Resident. Notes: *Creea, creea*, rendered by Mr. Seebohm as *khēē̆-ŏ, khĕŏ-ŏ*. Utters a quick, chattering noise at times, like a starling. Local and other names : Cornish Daw, Cornish Chough, Cornwall Kae, Market Jew, Chauk Daw, Red-legged Crow, Killigrew, Hermit Crow, Cliff Daw, Gesner's Wood Crow. Gregarious, but decreasing in numbers. Not a very close sitter, and noisy when intruded upon.

COOT.

Description of Parent Birds.—Length about eighteen inches. Bill of medium length, nearly straight, and dull white tinged with red. There is a smooth, naked white patch on the forehead which readily distinguishes this bird from the Waterhen, with its red shield. Irides hazel. Under the eye is a narrow, curved line of white. The whole of the plumage is black with exception of white on the bend of the wing and narrow bar formed by the white tips of the secondaries. The under-parts are tinged with bluish-grey. Legs, toes, and scallop-shaped lobes on either side of the latter, dark green. Above the knee is an orange-coloured garter.

The female is similar in size and appearance to the male.

Situation and Locality.—Amongst reeds, rushes, osiers, and other aquatic herbage in marshes, by the sides of ponds, reservoirs, large sluggish rivers, and lakes. They are generally built up from the bottom, but are sometimes simply moored to surrounding objects, and becoming detached by wind or floods, float about without any apparent

COOT.

inconvenience to the builder. I recollect once seeing an instance of this kind in a large reservoir in South Yorkshire. Breeds in suitable places in nearly all parts of the British Isles. Our illustration was procured in Norfolk.

Materials.—Decaying sedges, flags, reeds, and rushes, and although not very elegant is wonderfully strong. It is a very pretty sight to see this bird pulling up decayed weeds and swimming with them to its half-constructed nest.

Eggs.—Seven to ten. As many as fourteen or fifteen have been found. Dingy stone-colour or dull buff, spotted and speckled all over with nutmeg brown. Size about 2·1 by 1·5 in.

Time.—April, May, June, and July.

Remarks.—Resident, and partially migratory. Note, a clear, ringing *ko*. Local and other names: Common Coot, Bald Coot. Sits lightly.

CORMORANT.

Description of Parent Birds.—Length about thirty-six inches. Bill rather long, straight except at the tip, where it is hooked, and pale brown. Irides green. Crown, nape, and a portion of the neck black, intermixed with a number of very narrow white feathers almost like hairs. The feathers at the back of the head are elongated into a kind of crest. The feathers on the back and of the wing-coverts are dark brown bordered with black. Wing and tail quills black. Round the base of the bill and chin bare and yellow, bordered with white on the latter. Neck all round below the portion streaked with white, breast, belly, and under-parts rich bluish-black, except on either thigh, where there is a patch of white. Legs, toes, membranes, and claws black. The female is said not to be so large as the male, but has her crest often longer.

Situation and Locality.—On ledges of cliff by the sea, and also inland, on low rocky islands and reefs, sometimes in trees. The bird breeds pretty generally round our coasts wherever suitable cliffs and rocks are to be found, and has established inland colonies at several places. We were unfortunately unable to land on the Cormorant Rock at the Farne Islands on account of a heavy sea running, at the time we visited it in 1894. As we approached, the birds left their nests one by one and flew out to sea. The stench on the leeward side of the island was so intolerable that it turned my brother sick. In spite of this he, however, tried two flying shots from the boat; but the tide runs amongst the Farne group like a millrace, and we were soon carried out of focus. Next year (1895) we were determined to land upon the Cormorants' Rock (Megstone), and at length succeeded in doing so at considerable personal risk, but, alas! only to find that a recent gale had swept every nest and egg off it. We, however, took a photograph of the new nests the birds were busy building, as represented in our illustration.

Materials.—Sticks, twigs, and coarse grass or seaweed, depending upon locality. It is a large, high nest.

Eggs.—Three generally, but sometimes as many as five or six. The real shell is a pale blue, but this is usually hidden by a thick coating of chalk which can easily be scraped off. Sometimes the real shell shows through the casing of lime. Size about 2·6 by 1·62 in.

Time.—April, May, and June.

*Remarks.- -*Resident. Note : remarkably silent, but occasionally utters "a harsh croak" according to Seebohm. Local and other names, Crested Cormorant, Corvorant, Great Black Cormorant, Cole Goose, Sea Crow, Scart Brongie, Norie. Sits closely or lightly according to situation.

CRAKE, BAILLON'S.

A rare bird, whose nest and eggs have only been found about twice in the British Isles, and in neither case within recent times.

CRAKE, CORN.

Description of Parent Birds.—Length about ten and a half inches. Bill of medium length, nearly straight, thick at the base, and light brown. Irides hazel. Over the eyes and ear-coverts, also cheeks, ash-grey. Head, neck, back part of wings, tail-coverts, and feathers reddish-brown, with a long dusky streak in the centre of each. Wing-coverts rich bay, quills reddish-brown. Breast, belly, and under parts pale yellowish-brown or light buff, barred on the sides with two shades of reddish-brown. Legs, toes, and claws light yellowish-brown.

The female is rather inferior in size to the male, and the colours on the cheeks and wings less distinct.

Situation and Locality.—On the ground amongst mowing grass, clover, willow beds, and standing corn all over the United Kingdom.

CORMORANTS.

Materials.—Strong stems of dead grass and leaves, with an inner lining of finer grass.

Eggs.—Seven to ten; as many as twelve and even fifteen have been met with, however. Pale reddish-white or light buff, spotted, freckled, and blotched with red-brown of various shades and ash-grey. Size about 1·4 by 1·1 in.

Time.—May and June, although nests with eggs in have been reported as late as September.

Remarks.—Migratory, although individuals remain all the year round in Ireland. Arrives in April and May and departs in September and October, although stragglers remain later. Notes: *Crake, crake.* Local and other names: Landrail, Meadow Crake, Corn Creak, Draker Hen. The bird sits close, and, as a consequence, individuals sometimes get their heads cut off by the mower's scythe or machine.

CRAKE, SPOTTED.

Description of Parent Bird.—Length about nine inches. Bill of medium length, straight, thick at the base and pointed at the tip, yellowish-brown in colour, with a brighter and redder tinge towards the base. Irides brownish-hazel. Crown hazel-brown spotted with black in the middle; over each eye is a patch of dull blue-grey; sides of head, nape and sides of neck olive-brown spotted with white. The feathers of the back are black, broadly edged with dark olive-brown, and streaked up and down with fine lines of white. Wing-coverts olive-brown; quills dark brown, mottled and barred with white; rump, upper tail-coverts, and tail-quills black, bordered with dark reddish-brown and spotted with white. Chin grey-brown; throat and breast dusky brown mottled with white; sides and flanks greyish-brown barred with white; under tail-coverts buffish-white, legs and toes yellowish-green (toes long), claws brown.

The female is a trifle smaller, and not quite so distinctive in coloration.

Situation and Locality.—In a tussock of sedge, amongst reeds and other vegetation growing in marshes, bogs, and wet, swampy ground. The foundation is generally in water. Sparingly along the East Coast counties, in Wales, Cumberland, one or two suitable parts of Scotland, and Ireland.

Materials.—Coarse aquatic plants, such as reeds and flags in somewhat liberal quantities, and lined with dry grass.

Eggs.—Seven to twelve. White, yellowish-grey, or ocherous. Some authorities describe them as being occasionally white tinged with green, or grey tinged with pink, spotted with dark reddish-brown, and underlying markings of grey. Size about 1·3 by ·9 in.

Time.—May is the principal laying month; however, eggs have been found in April and right through June.

Remarks.—Migratory and resident. The first kind of birds arrive in March and depart in October. Notes: *whuit, whuit.* Local and other names: Spotted Rail, Water Rail, Water Crake (this name is also applied to the Dipper), Spotted Gallinule, Spotted Water Hen. Slips quietly off nest and hides amongst surrounding vegetation.

CREEPER, TREE.

Description of Parent Birds. — Length about five inches. Bill rather long, curved downwards, dark brown on the top, and dirty white, tinged with yellow, underneath. Irides hazel. Crown dark brown, spotted and streaked with pale brown. Back of neck, back, and rump tawny-brown, mixed with ash-grey. Wing-coverts brown, tipped with greyish-yellow; quills variegated with brown and black, some of them tipped with light grey. Tail tawny-brown, the feathers being strong in the shaft, and, from their help to the bird in climbing and holding on to the bark of trees, often worn quite bare at the ends. Chin, throat, breast, and belly greyish-white, inclining to rusty reddish-white on the flanks and vent. Legs, toes, and claws, which are very long, light brown.

The female is similar in size and colour to the male.

Situation and Locality.—In a hole in a tree; behind a loose piece of bark still clinging to a decayed tree; amongst piles of stacked timber; in niches and crevices of buildings, and behind half-detached pieces of plaster. The one in our illustration was placed behind a sound piece of the outer shell of a decayed pollard. The bird could either slip off from the front, or up a kind of chimney, having its exit just under the face of the inquisitive onlooker. In nearly all well-wooded districts throughout the British Isles.

Materials.—Fine twigs, dead grass, sometimes little chips of decayed wood, wool, moss, feathers, and rabbits' down.

Eggs.—Six to nine, white, spotted and speckled with reddish-brown and sometimes dullish purple

SITE OF TREE CREEPER'S NEST.

D

spots, generally in a kind of zone round the larger end. Occasionally more distributed. Size about ·66 by ·47 in.

Time.—April, May, and June.

Remarks.—Resident. Notes: song not often heard, but high, shrill, and not unpleasant. Local and other names: Creeper, Tree Climber, Common Creeper. Sits very closely.

CROSSBILL.

Description of Parent Birds.—Length from six and a half to seven inches. Beak rather large, upper mandible turned down and lower up. They do not lie in consequence in a straight line over each other, but cross like the blades of a pair of scissors. The Crossbill varies more according to age, sex, and individual than perhaps any other British bird.

Swaysland gives the following description: "When young the male birds are greenish-brown, with a tinge of olive, the whole being speckled with darker brown; they are, however, lighter upon the under-parts; but after the first moult a red tinge prevails, occasioned by the tipping of the feathers with that hue. The red is much darker upon the upper-parts. At the second moulting these colours are lost, and the bird's plumage becomes an olive-brown. shaded over with greenish-yellow upon the back, though it is much lighter upon the under-parts, and is speckled with orange upon the breast and rump.

"The females are, however, either grey with a little green on the head, breast, and rump, or else speckled in an irregular manner with those colours."

Professor Newton, in describing the male with the second dress on, says: "A red male that had completed his first autumnal moult had the bill dull reddish-brown, darkest towards the tip of the upper mandible; irides dark brown; the head, rump, throat, breast, and belly tile-red; the feathers on the back mixed with brown, producing a chestnut brown; wing and tail feathers nearly uniform dark brown; vent and lower tail-coverts greyish-white; legs, toes, and claws dark brown. . . . Young females, after their first striated dress, acquire a greenish-yellow tint on the crown and the lower parts of the body mixed with greyish-brown; the rump and upper tail-coverts of primrose yellow tinged with green; wings, tail, and legs as in the male."

A celebrated Continental authority, writing upon the matter, says: "If the Crossbills are grey or speckled, they are young; if red, they are one year old and have just moulted; if carmine, they are just about to moult for the second time; if spotted with red and yellow, they are two years old and in full feather."

Situation and Locality.—On the branches of Scotch and other fir-trees, sometimes quite close to the bole or stem, at others some distance away on a horizontal branch at varying heights from the ground. Generally in plantations of cone-bearing trees over the greater portion of England, Scotland, and in Ireland, where suitable plantations are to be met with. The bird is very sparsely distributed, and uncertain in its patronisation of recognised breeding haunts. It is most numerous in the central counties of Scotland.

Materials.—Twigs, roots, coarse dead grass, lined internally with finer grass, hairs, and feathers. It is similar in construction and appearance to that of the Greenfinch.

Eggs.—Four to five. White, sometimes faintly tinged with pale blue, very sparingly speckled with reddish-brown and pale brown. Average size about ·9 by ·67 in.

Time.—Some English authorities say February and March, others February, March, and April, and Continental authorities December to April.

Remarks.—A winter visitor, but a few pairs stay to breed. Notes: call, *chip-chip-chip*. Other notes used whilst flying from tree to tree, *soc-soc-soc*. Local and other names: Common Crossbill, Shell Apple, European Crossbill. A very close sitter.

CROW, CARRION.

Description of Parent Birds.—Length about eighteen inches. Bill fairly long, strong, and black. The base of the beak is covered with bristles, which stand forward. These bristles and its hoarser note distinguish it from the Rook. Irides dusky. The whole of the plumage is black, glossed above with a lustrous greenish sheen. Legs, toes, and claws black.

The female is about the same size, but lacks a little of the metallic lustre which characterises the male.

Situation and Locality.—In high trees, generally on a large branch near the bole, and at a good height from the ground, on the outskirts of wood and plantations; sometimes on ledges of cliffs. In England, Scotland, and the North of Ireland. It is nowhere very numerous, as its predatory habits make for it an uncompromising

CARRION CROW.

enemy in the gamekeeper. Our illustration is from a photograph taken in Westmoreland.

Materials.—Sticks and mud, lined with grass, wool, horse and cowhair.

Eggs.—Four to five, occasionally as many as six, grey-green, blotched and spotted with ash-colour or smoky-brown ; sometimes they are found quite blue, and minus spots of any kind. They are similar to those of the Rook and Raven, but larger than the former and smaller than the latter, and the position of the nest generally suffices to distinguish them. Size about 1·65 by 1·2 in.

Time.—April and May.

Remarks.—Resident. Notes, a hoarse croak. Local and other names, Crow Mussel (from its habit of eating mussels), Doup, Gor Crow, Minden Crow, Black-nebbed Crow. Sits lightly, and generally in such a position as to command a good surrounding view.

CROW, GREY. *See* CROW, HOODED.

CROW, HOODED. *Also* GREY CROW *and* ROYSTON CROW.

Description of Parent Birds. — Length about twenty inches. Beak moderately long, pointed, strong, and black ; the base is covered by stiff projecting feathers. Irides dusky. Head, throat, wings, and tail a shining blue-black. Nape, back, rump, and under-parts generally dark slaty-grey. Legs, toes, and claws a shining black.

The female is a little smaller in size, and the slaty-grey parts of her plumage are tinged with brown.

HOODED CROW.

Situation and Locality.—Rocks, cliffs, and trees. An instance has been recorded of this bird's building on the roof of a crofter's hut. In Ireland, the mainland of Scotland, and the islands to the west and north. Our illustration was obtained on a small rocky island near Oban. The young ones were ready to fly, and indeed did so on being disturbed. The day was very hot, and distressed them until they all gasped like hard-run spaniels.

Materials. — Sticks, twigs, heather, and ling, with an inner lining of roots, moss. wool, hair or feathers.

Eggs.—Three to six : generally five, grey-green in ground colour, blotched and spotted with varying shades of olive or greenish-brown. Variable both in regard to size, shape, ground colour, and colour of markings. Size about 1·65 by 1·2 in.

Time.—March, April, and May.

Remarks. — Resident, but subject to southern movement in winter. Note, a hoarse croak. Local and other names : Hoody, Dun Crow, Grey Crow, Bunting Crow, Royston Crow, Greyback, Norway Crow. Kentish Crow, Scarecrow. The bird is a light sitter, and often interbreeds with the Carrion Crow.

CROW, ROYSTON. *See* CROW, HOODED.

CUCKOO.

Description of Parent Birds. — Length about fourteen inches. Bill rather short, slightly curved downwards, and black, turning yellowish at the base. Irides yellow. Head, nape, back, and upper parts generally dark ash colour. Wing-quills dusky, barred with white for some distance on their inner webs. Tail-quills greyish-black, especially the middle feathers, tipped with white, and marked with white spots. Chin, throat, and upper breast pale ash-grey; lower breast, belly, vent, and under tail-coverts white, marked with wavy, transverse bars of black; the two last parts often have a reddish-brown tinge. Legs, toes, and claws yellow.

The female is very similar in appearance to the male, but a little smaller in size.

Situation and Locality. — Deposits generally a single egg in the nest of the Meadow Pipit, Pied Wagtail, Grey Wagtail, Hedge Sparrow, Sedge Warbler, White-throat, Robin, Yellow Hammer, Jay, and Martin, although the last two must be rarely patronised.

Our illustration is from a photograph of a Pied Wagtail's nest, containing a Cuckoo's egg, which could only be distinguished by its greater size and shape and, on being blown, its thicker shell. The nest was situated about nine feet from the ground, amongst ivy growing over a high garden wall. A common summer visitor to all parts of the British Isles. I have noticed that in the more elevated parts of the North of England, Meadow Pipits rear more young Cuckoos than all the other foster-parents put together.

Materials.—None.

Eggs.—It is certain that the bird lays more than one egg; but although naturalists of good repute have mentioned the number as five, and others have been of opinion that even a larger number may be laid, there is, so far as I know, no reliable evidence to support either supposition. I have never noticed that young cuckoos exceeded in numbers the old ones, in a given district, where

I was out of doors all day long, every day in the year, and many years together, in unbroken succession. But, of course, the number hatched could never represent the number laid, although the place to which I refer was singularly free from vermin and collectors. A single egg is found in the nest of a foster-parent, but occasionally two have been seen, and in very rare instances even three ; but it is, of course, impossible to say whether they were laid by the same bird. The egg of the Cuckoo is small in size compared with its layer, and varies very much in coloration, but, strangely enough, often harmonises closely with those of the bird in whose nest it is deposited. It is usually reddish-grey, mottled and spotted closely, with darker markings of the same colour, or pale greyish-green, marked with spots of the same colour. Size about ·87 by ·75 in.

Time.—April, May, and June.

Remarks. — Migratory, arriving in April and leaving in July, the young ones in August and September. Notes : song, *cuckoo*, and a churring note. Local and other names : Gowk, Common Cuckoo. The young Cuckoo turns out all the other members of the nest in which it is hatched, an operation to which I was witness on one occasion.

CURLEW, COMMON.

Description of Parent Birds.—Length varying from twenty-one to twenty-two inches. Bill very long, slender, curved downwards, and dark-brown, paler at the base of the under mandible. Irides hazel. Head, neck, upper part of back, scapulars, and wing-coverts pale brown, with a dark brown

CUCKOO'S EGG IN PIED WAGTAIL'S NEST.
(The one on the left is the Cuckoo's Egg.)

streak in the centre of each feather. Wing-quills black, spotted and marked with light brown on the inner webs. Lower back and rump white, marked by a few dusky spots. Upper tail-coverts white, marked with dark brown; tail-feathers barred with dull yellowish-white and dark brown. Chin white; throat and upper part of breast very pale brown, marked with dark brown streaks; lower part of breast, belly, vent, and under tail-coverts white, spotted on the two first, with blackish-brown and a dusky streak or two on the latter. Legs long, and, like the toes, bluish-grey in colour.

The female is similar in plumage, but is larger, sometimes even to the extent of five inches in length.

Situation and Locality.—On the ground amongst long, coarse grass, tufts of rushes and heath; sometimes quite exposed on bare ground. On rough, undrained pasture land, moors, and uplands in the West and North of England, Wales, Scotland, and Ireland. Our illustration is from a photograph taken on the Westmoreland Hills, where these birds are very common. We found a couple of nests within a few yards of each other, the one containing two, and the other three eggs; and the specimen figuring on the page opposite was only just over the wall in an adjoining pasture.

Materials. — A few short bits of dead rushes, withered grass, or dead leaves, placed in some small declivity; sometimes nothing whatever.

Eggs. — Four, sometimes only three, varying from olive-green to brownish-buff in ground colour, spotted and blotched with dark green and blackish-brown. Size about 2·65 by 1·85 in.

Time.—April, May, and beginning of June sometimes.

COMMON CURLEW.

Remarks.—Resident, but resorting to the coast-line during winter. Notes: *curlew, curlew,* uttered something like *gurleck, gurleck,* when the bird is alarmed. Local and other names: Whaap, or Whaup, Stock Whaap. A very light sitter.

CURLEW, STONE. *Also* Norfolk Plover, Great Plover, *and* Thicknee.

Description of Parent Birds.—Length about seventeen inches. Bill, short compared with that of the Common Curlew, strong, nearly straight, greenish-yellow at the base, and black at the tip. Irides golden-yellow. A light-coloured streak runs from the base of the beak, under the eye, to the ear-coverts, followed by a brown one running in the same direction below it. Crown, nape, and back of neck light brown, streaked with black. Back, wings (except primaries, which are nearly black, with a white patch on the end of the first and second feathers), and upper tail-coverts light brown, each feather having an elongated blackish-brown centre. Upper half of tail-quills of two shades of brown, producing a mottled effect, followed by a band of white and a black tip. Chin and throat white; front of neck and breast very light brown, streaked with blackish-brown; belly and sides nearly white, streaked with brown. Vent and under tail-coverts creamy-white, unmarked. Legs and toes yellow; claws black.

The female is very much like the male in her plumage.

Situation and Locality.—On the ground in warrens, on downs, heaths, and dry commons, principally in Norfolk and Suffolk, but found in

STONE CURLEW

several other counties, as far west as Dorsetshire
and as far north as Yorkshire. Our illustration
was procured on a common in Norfolk.

Materials.—Sometimes a few bits of grass, but
generally nothing whatever, in the slight declivity
made or selected.

Eggs.—Two, varying in ground-colour from
greyish-yellow to clay-colour, blotched, spotted,
and streaked with dark brown, light brown, and
greyish-blue. They harmonise very closely with
their surroundings. Size about 2·1 by 1·55 in.

Time.—May and June. Eggs have, however,
been found as late as September.

Remarks. — Migratory, arriving in April and
departing in October or November. Note very
loud and shrill, and uttered particularly at dusk of
evening. Local and other names : Norfolk Plover,
Stone Plover, Thicknee, Common Thicknee, Thick-
kneed Bustard, Whistling Plover. A light sitter.

DABCHICK. *See* GREBE, LITTLE.

DAW. *See* JACKDAW.

DIPPER. *Also* WATER OUZEL.

Description of Parent Birds.—Length about
seven and a half inches. Bill of medium length,
nearly straight, and black. Irides hazel. Head
and back of neck dark brown. Back, wings, rump,
and tail, which is short, black. Chin, throat, and
upper breast snowy white. Belly chestnut brown
or rust colour, vent and under tail-coverts black.
Legs, toes, and claws black.

DIPPER'S NEST.

E

Situation and Locality.—I have found this bird's nest in niches of rock, on sloping ledges, on a boulder in the middle of a mountain stream, behind a waterfall, in the root of a tree, in the arch of a bridge where a stone had slipped out, fixed to a sod which was constantly dripping with splashes from a waterfall close by, and on one occasion in a tree some twelve feet over a stream. It is never far from a mountain torrent. It is met with in the West and North of England, Wales, and pretty generally over Scotland and Ireland. Our illustration is from a photograph procured on the upper Eden, and was taken with the camera standing in a rushing torrent.

Materials.—The exterior is made of aquatic mosses, generally harmonising closely with surrounding objects, and the inside is beautifully lined with dead leaves laid layer upon layer. The appearance of the nest varies considerably according to situation. Our illustration represents rather a squat one.

Eggs.—Four to six, generally five, of a delicate semi-transparent white, unspotted. Size about 1·0 by ·75 in.

Time.—March, April, May, June, and occasionally as late as July. I used to notice when a boy that on some seasons the Thrush, and on others the Dipper, would commence to nest first in the North of England.

Remarks.—Resident. Call notes: *chit, chit.* Song low and sweet, but very pleasant, and uttered right through the hard frosts of winter during gleams of sunshine. Local and other names: Water Ouzel, Bessy Dooker, Brook Ouzel, Water Crow, Water Piet, Water Crake. Sits closely, and when suddenly disturbed will often dive into a near-by pool.

DIVER, BLACK-THROATED.

Description of Parent Birds.—Length about twenty-six inches. Bill rather long, straight, pointed, and black. Irides red. Top of head and back of neck grey, darkest on the fore part of head. Sides of neck marked with longitudinal black and white lines. Back scapulars and wing-coverts black, the two first being marked with square patches and the last with round spots of white. Wing-quills, rump and tail feathers, dusky black. Chin and throat black, divided by a collar of black and white short longitudinal lines. Breast, belly, and vent white. Under tail-coverts dusky. Legs, toes, and webs dark brown on the outside, reddish or pale brown on the inside.

The female is a trifle smaller than the male.

Situation and Locality.—In a hollow on the ground, amongst the stones and shingle of secluded mountain tarns and loch shores, sometimes amongst the grass, but rarely far from the water ; also on small grassy islands in bodies of fresh water. The bird is a great lover of old haunts. In the North-west of Scotland and the outer Hebrides.

Materials.—Roots, stalks, or aquatic herbage, lined with grass, sometimes nothing whatever.

Eggs.—Two, occasionally only one, ranging in colour from buffish-brown to dark olive-brown, scantily spotted with umber and blackish-brown. Average size about 3·25 by 2·0 in.

Time.—May and June.

Remarks.—Resident, but subject to southern movement in winter. Notes strange and weird, and said to resemble " *Drink ! drink ! drink '* the lake is nearly dried up." Local and other names Speckled Loon, Lumme, Northern Douker. Does not sit closely.

DIVER, RED-THROATED.

Description of Parent Birds. — Length about twenty-four inches. Bill rather long, straight, sharp-pointed, and bluish horn-colour. Irides red. Face and crown ash-grey, nape nearly black, with short perpendicular white lines on it. Back, wings, and upper tail-coverts almost black, spotted with white, except wing-quills, which are uniform black. Chin, cheeks, and sides of neck ash-grey, mixed with lines and spots of a lighter tinge. On the upper part of the neck in front is a conical patch of chestnut-red. Breast, belly, vent, and under tail-coverts white ; sides greyish-black, spotted with white. Legs, toes, and webs dark brown on the outside and lighter within.

The female is somewhat smaller in size than the male.

Situation and Locality.—On the turf or amongst stones and shingle close to the edges of lonely moorland and mountain pools, tarns, or lakes ; by preference on a small island in any of the above sheets of water, on the northern and western mainland, and the islands off those coasts of Scotland. The bird is also said to breed in the west of Ireland. Our illustration was procured in the outer Hebrides.

Materials. — Loose rushes and dry grass, very often nothing at all.

Eggs.—Two. Olive, or deep greenish-brown in ground colour, spotted with blackish-brown. Size about 2·8 by 1·8 in.

RED-THROATED DIVER.

Time.—May and June.

Remarks.—Resident, but subject to much local movement. Notes, *kakera*, *kakera*, uttered during the breeding season. Local and other names : Rain Goose, Kakera, Cobble, Speckled Diver, Spratoon, Sprat-borer. Not a very close sitter.

DOTTEREL.

Description of Parent Birds.—Length about nine and a half inches. Bill of medium length, straight and black. Irides dark brown. Crown and nape blackish-brown. A broad white line runs from the base of the beak, over the eye, down behind the ear-coverts, which are ash-coloured, as also are the neck and back. Wings ash-brown, except quills, which are ash-grey. Tail olive-brown tipped with white; chin white; throat and sides of neck grey. A white gorget-like band, bordered on either side by a dark line, runs across the breast and ends at each shoulder. Breast pale, dull orange; belly black; vent and under tail-coverts white, slightly tinged with buff. Legs and toes greenish-yellow; claws black.

The female is a little larger, and more handsomely marked.

Situation and Locality.—On the ground amongst woolly-fringe moss and other coarse mountain vegetation, on high, wild moorland districts and lonely mountains of Scotland. It used to breed in the Lake District until recently, but I think the greed of the fly-fisher has for ever sealed its doom so far as England is concerned. I am myself an ardent fly-fisher, and have been offered handsome sums of money in my artificial fly-dressing days if I would only procure the skin of this bird, whose feathers are popularly supposed to exercise a kind of charm over trout in some northern districts.

Materials.—None. The eggs are simply deposited in a slight declivity trodden in the place selected.

Eggs.—Three, yellowish-olive to dark cream in ground colour, thickly blotched and spotted with dark brown or brownish-black. Size about 1·65 by 1·15 in.

Time.—May, June, and July.

Remarks.—Migratory, arriving in April and May and leaving in August and September. Notes, "*durrdroo.*" Local and other names, Foolish Dotterel, Dotterel Plover. Gregarious; very tame and stupid; sits closely, and resorts to the usual deceit of many other ground builders in order to get rid of the intruder.

DOTTEREL, RINGED. *See* PLOVER, RINGED.

- -

DOVE, RING.

Description of Parent Birds.—Length about seventeen inches. Bill moderately long, curved downwards slightly at the tip and pale red, yellowish towards the end and whitish on the soft parts surrounding the nostrils. Irides straw colour. Head and upper part of neck bluish-ash colour. The sides of the lower part of the neck are beautifully glossed with green and purple, according to the light upon them. On either side of the neck is a patch of glossy white, which almost meets at the back. Back and wing-coverts bluish-grey, with exception of a few feathers of the latter, which are white, and during flight form a conspicuous patch, by which the bird may easily be distinguished from any other member of the pigeon family. Quills dark grey edged with white; the feathers of the spurious wing are almost black. Tail-quills of varying shades of grey, darkest towards the tip. Chin bluish-grey; neck and breast glossy purple and green; belly and under-parts light ash-grey. Legs and toes red; claws brown.

The female is somewhat duller in plumage and smaller in size.

Situation and Locality.—In fir, yew, whitethorn, and various other kinds of trees. I have met with it on the crown of a pollard, and frequently on the ivy-clad trunk of a tree, growing almost at right angles from high rocks and precipices. I have also met with it upon several occasions in an isolated thorn bush growing in the middle of a large field. Our picture is from a photograph, to obtain which it was necessary to tie the camera into a high tree. The nest is situated at a height of from five to seventy or eighty feet, and is found all over the United Kingdom where suitable woodland is to be met with.

Materials.—Dead twigs and sticks woven into a loose platform. The nest is often such a poor, flimsy affair that the eggs may be seen through it from beneath, and it is frequently blown down by gales of wind. On the other hand, I have on one or two occasions met with nests of a substantial character, into which bits of dried sods had been introduced, so that not even a ray of light could find its way through.

Eggs.—Two. White and glossy, similar to those of the Rock Dove, but larger. Average size about 1·65 by 1·25 in.

RING DOVE.

Time.—March, April, May, June, and July. However, nests have been found in nearly every month of the year.

Remarks.—Resident. Notes, a soft *coo-coo-co-co, coo.* Local and other names: Cushat, Wood Pigeon, Quest, Cushie Doo, Ring Pigeon. Sits pretty close when well hidden, but lightly when in an exposed situation.

DOVE, ROCK.

Description of Parent Birds. — Length about fourteen inches. Bill of medium length, nearly straight, and brown tinged with red. Irides light reddish-orange. Head and neck dark bluish-grey, glossed with a purply-red and green sheen on the latter; back and wing-coverts light pearl-grey; greater coverts and secondaries marked with black bars; quills bluish-grey, darkest towards the tips; rump white; tail-coverts bluish-grey; quills bluish-grey, with a darker bar at the tips; chin dark bluish-grey; throat and upper part of breast glossed with green, lavender, purple, and purplish-red; lower breast and all under parts grey. Legs and toes purplish-red; claws brown.

The female is somewhat smaller, and not so brilliant and distinctive in her coloration.

Situation and Locality.—Ledges and clefts of maritime and inland cliffs, generally the former, round the coasts of England, Wales, Scotland, and Ireland, wherever suitable accommodation is to be met with. Our illustration represents a cave in North Uist where numbers of these birds and shags were breeding together.

Materials.—A small collection of twigs, sticks, seaweed, and bents, roughly constructed, and flat.

CAVE IN WHICH NUMBERS OF ROCK DOVES AND SHAG
WERE BREEDING.

Eggs.—Two, white, unspotted and smooth. Size about 1·15 by 1·15 in

Time.—March, April, May, and June, although eggs have been found in nearly every other month of the year.

Remarks—Resident. Notes, *coo-roo-coo*, last syllable prolonged. Local and other names: Rockier, Wild Pigeon, Rock Pigeon, Wild Dove, Doo. A fairly close sitter, and distinguished from the Stock Dove by its white rump.

DOVE, STOCK.

Description of Parent Birds.—Length about thirteen and a half inches. Bill of medium length, nearly straight, and pale red, whitish at the tip. Irides brown. Head, neck and upper parts of back deep bluish-grey, glossed on the sides of the neck with green and purplish-red. Wing-coverts bluish-grey, spotted and marked with black on the greater ones; quills brownish-grey, turning bluer towards the tips. Rump and upper tail-coverts pale bluish-grey. Tail bluish-grey for about two-thirds of its entire length, then crossed by a band of lighter ash-grey, and the end, which is rounded, of so dark a grey that it may almost be described as black; the exterior webs of the outside feathers nearly white. Breast pale reddish-purple; belly, thighs, and under tail-coverts ash-grey; legs and toes red.

The female is a little smaller, and her markings are not so well defined.

Situation and Locality.—In hollow trees, rabbit-holes, crevices of rock; in quarries and cliffs, old Crow or Magpie's nests; in ivy growing against

CLIFF IN WHICH NUMBERS OF STOCK DOVES BREED.

trees, rocks, towers, and steeples; and sometimes quite on the ground, under dense furze bushes. Our illustration represents a much-frequented cliff in Westmoreland during the breeding season. The bird is a common breeder in the eastern and midland counties, and many other parts of England and Wales, and has gained a footing in both Scotland and Ireland, according to Mr. Dixon.

Materials.—Twigs, roots, and straws in small quantities, and arranged with very little care or skill.

Eggs.—Two, white, faintly tinged with cream colour. They are smaller than those of the Ring Dove, and the creamy tinge distinguishes them from those of the Rock Dove. Size about 1·45 by 1·15 in.

Time.—February to October.

Remarks.—Resident. Notes, *coo-oo—oo*, the last syllable longer than the first. Local and other names, Stock Pigeon, Wood Dove, Wood Pigeon. Sits closely. Gregarious, as a rule. It may easily be distinguished from the Rock Dove by its lack of a white rump.

DOVE, TURTLE.

Description of Parent Birds.—Length about twelve inches. Bill of medium length, slightly curved downward at the tip, and brown. Irides reddish-brown. Crown and back of neck ash-grey, mixed with olive brown. Back, part of wings, and rump ash-brown, lightest on the margins of the feathers. Wing-quills dusky brown with lighter margins and tips. Tail-coverts dusky brown, quills the same in the centre, rest dark grey tipped with white, with which the outside feathers on either side are margined. Chin pale brown; throat and upper breast light purplish-red, fading into grey. The sides of the neck are marked with a patch of black, each feather of which is tipped with white. Belly, vent, and under tail-coverts white. Under-side of tail-feathers black, deeply tipped with white, except two centre ones, which are of a uniform dusky brown.

The female is rather smaller in size, lacks the black feathers tipped with white on the sides of the neck, and is duller and less distinctive in her coloration.

Situation and Locality.—In tall, rough hedges, whitethorn and holly bushes; in woods, plantations, copses, and spinneys. Common in the Southern, Midland, and Eastern Counties; scarcer in the West and North; not met with as a breeder in Scotland, and only sparingly in Ireland. Our illustration is from a photograph taken in Surrey. I have found five or six nests in little more than a couple of acres of wood. The nest is placed at a height of from four or five to twenty feet.

Materials.—Sticks and twigs carelessly made into a slight platform, through which the eggs may invariably be seen from beneath. I remember once finding one amongst some tall ash saplings, that was so slightly constructed with birch twigs as to endanger the eggs slipping through it. I have, however, on the other hand, a nest before me as I write which is a thick and bulky platform made entirely of roots of weeds collected from an adjoining ploughed field.

Eggs.—Two, creamy-white, glossy, oval, and unspotted. Size about 1·18 by ·90 in.

Time.—May, June, July, and August.

Remarks.—Migratory, arriving in April and May,

and departing in September and later. Note, *tur-tur*, repeated rapidly. Local and other names: Wrekin Dove, Ring-necked Turtle, Common Turtle. Sits pretty closely when incubation has advanced, and when disturbed, flies off without demonstration.

— —— — ——

DUCK, EIDER.

Description of Parent Birds.—Length about twenty-five inches. Bill moderately long, thick at the base, straight, and dirty green, whitish at the tip. Irides brown. The top of the head, including the parts round and on a level with the eyes, velvety black, turning to a palish green on the ear-coverts and back of the head. Neck, back, wing-coverts, and scapulars white. Some of the coverts are elongated, and curved at the ends falling over the quills; greater coverts, and quills black. Rump black, tail feathers brownish-black. The lower part of the neck is white, tinged with buff. Breast, belly, and under-parts black; flanks patched with white. Legs, toes, and webs dusky-green: claws dusky.

The female is somewhat smaller, and pale reddish-brown, variegated with brownish black.

Situation and Locality.—On the ground amongst coarse grass, in clefts of rock, and sometimes in collections of seaweed, amongst shingle, on rocky islands, and on the coast at suitable places round Scotland and at the Farne Islands. Our first illustration represents an Eider Duck sitting on her eggs, and the second a nest containing only three eggs, and yet with down in it. We stroked the back of one bird as she sat on her eggs close under the walls of St. Cuthbert's Tower. The

TURTLE DOVE.

F

keepers told us the same bird had nested in that situation for seven years in succession, as they were able to identify her by a white spot at the back of her head.

Materials. — Dry seaweed, heather, or coarse grass, with an inner lining of beautiful soft down from the bird's own body. The down is accumulated as the eggs are laid or incubation advances. Individual birds vary in respect to supplying it during the time they are laying, for at the Farne Islands I noticed that whilst some had a fairly liberal supply and only three eggs, others with seven had not a particle.

Eggs.—Five to eight. Pale greyish-green to grey-cream colour, smooth and unspotted. Size about 3·0 by 2·0 in.

Time.—May and June.

Remarks. — Resident, with a more southern range in winter. Note : a harsh *kr, kr, kr.* Love note of male : *ah-oo.* Local and other names : St. Cuthbert's Duck, Common Eider, Colk, Dunter Duck. Sits very closely indeed.

DUCK, PINTAIL.

Description of Parent Birds.—Length about twenty-seven inches, several of which are accounted for by the abnormally long tail. Bill moderately long, nearly straight, and dusky-black, leaden-grey on the sides. Irides dark brown. Head and upper part of neck in front, nape, and all back of neck dark reddish-brown. The sides and back of head are glossed with purple. Back, wing shoulders, and parts in front of them, grey ; wing tertials elongated, black in the centre, and bordered with white and grey ; greater coverts ash-brown tipped

EIDER DUCK SITTING ON HER NEST.

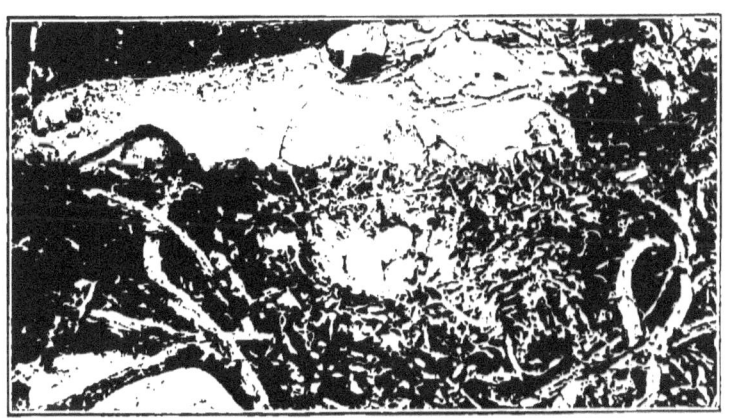

EIDER DUCK'S NEST.

with reddish-buff and white. Secondaries black, the outer web of each forming a patch of dark green; primaries greyish-brown. Tail-coverts ash-grey, elongated narrow-pointed feathers black, rest dark brown bordered with white. From the side of the head level with the eye a white streak runs down each side of the neck, widening gradually as it descends until the middle of the neck is reached, from thence it opens out round the front of the neck, breast, and belly. Sides grey; vent and under tail-coverts black. Legs, toes, and webs blackish-brown.

The female is somewhat smaller; and her plumage is made up of varying shades of brown, the darkest colours in the centre of each feather and on the upper parts of the body. During July, August, and September the male assumes the dress of the female.

Situation and Locality.—On the ground amongst coarse grass, rushes, and similar herbage growing near ponds, lakes, and arms of the sea. It is a very rare breeder indeed in our islands, having only been reported from two or three quarters of Ireland and one or two in Scotland and the Hebrides.

Materials.—Reeds, grass, and other kinds of dead vegetation, lined with brownish tufts of down, faintly tipped with white, from the bird's own body.

Eggs.—Six to ten, usually seven or eight, smooth, rather elongated, and "greenish-white," according to Mr. Saunders, and "pale buffish-green," according to Mr. Dixon, in colour. Size about 2·15 by 1·55 in.

Time.—May.

Remarks.—Migratory as a rule, wintering with us and spending the summer in Iceland. It is probable that the few staying to breed in our islands have a farther southern range than those from Iceland. Note: a soft inward *quack*. Local and other names: Winter Duck, Cracker, Sea Pheasant. A close sitter, and has bred repeatedly in confinement.

DUCK, TUFTED.

Description of Parent Birds.—Length about seventeen inches. Bill of medium length, but little broadened towards the point, and bluish-grey with a black tip. Irides rich dark yellow. Head and neck glossy purplish-black. The feathers at the back of the head are narrow and elongated, and form a tuft or crest. Back, rump, wings, and tail black with a white bar running across the secondaries, which are tipped with black. There is a small spot of white on the chin. Breast, belly, and sides white. Vent and under tail-coverts black. Legs, toes, and webs dusky black.

The female has all those parts which are black in the male a dusky brown, and the white parts dirty grey, marked with irregular lines on the sides and flanks. She lacks the crest, or only enjoys it in a very modified form.

Situation and Locality.—In a tuft or bush of long, coarse vegetation, such as rushes, sedges, heather, or bent grass, on the edges of tarns and lakes and other suitable places throughout the British Isles. I have met with odd pairs nesting round tarns on the North Yorkshire moors. It nests most numerously in Nottinghamshire. Our illustration was procured on a celebrated Norfolk Mere, and was in an ideal position, the bird having not more than twelve or fourteen inches to travel from the edge of her nest into deep water.

Materials.—Rushes, sedges, reeds, and grass, with an inner lining of down tufts plucked from the bird's own body. These are greyish-black, smaller and a trifle darker than those of the Pochard, with more obscure white centres.

Eggs.—Eight to fourteen, usually nine or ten. Pale buff tinged with green. Very similar to those of the Pochard. Size about 2·3 by 1·6 in.

Time.—May and June.

Remarks.—A winter visitor, though numbers stay to breed. Notes : call, *currugh*, *currugh*, uttered on alighting. Local and other names : Tufted Pochard. Sits closely.

DUCK, WILD. *Also* MALLARD.

Description of Parent Birds.—Length about twenty-four inches. Bill of medium length, broad, and yellowish-green. Irides hazel. Head and upper half of neck rich glossy green, below which is a narrow collar of white, succeeded by greyish chestnut-brown ; back brown ; wings ash-brown, with a broad transverse bar of reflecting purplish- or violet-blue, bounded on either side by a narrow bar of rich black, and another beyond of white. Rump, upper tail-coverts, and four middle tail-feathers, which are curled upwards, rich velvet-black, the rest ash-grey edged with white. Upper part of breast rich dark chestnut ; lower breast, belly, and vent greyish-white ; under tail-coverts rich black. Legs, toes, and webs orange-yellow.

The female is about two inches shorter, and her plumage is nearly all composed of sober brown and black. She retains the rich bar of violet- or

TUFTED DUCK.

purplish-blue on her wings, but lacks the curled feathers in the tail of the male. About the end of May the male commences to cast his curled tail-feathers and change his plumage, and during June he assumes the sober female garb, which he wears through July. This he begins to discard in August, and between the first and second weeks in October he has again donned his magnificent dress.

Situation and Locality.—On the ground amongst rushes, brambles, long rank grass, sedge tufts, under a bunch of heather, and in corn fields and hedge-bottoms. Generally near lakes, rivers, tarns, and ponds, or in marshes, bogs, and swamps. They are, however, often found at considerable heights, in faggot stacks, deserted Crows' nests, squirrel dreys, Hawks' nests, in hollow trees, pollards, and other elevated situations in ruins and rocks, from which heights the female conveys her progeny upon her back. Pretty generally in all suitable places throughout the British Isles. Our illustration is from a photograph taken in Essex.

Materials.—Dry grass, bracken, or other suitable vegetation near at hand, with a lining of down from the bird's own body. The tufts are neutral grey, tipped very slightly with white.

Eggs.—Eight to fifteen or sixteen, generally ten to twelve ; greenish-white tinged with buff. Size about 2·3 by 1·6 in.

Time.—February, March, April, May, June, and even as late as November, individual nests have been recorded. April and May are the principal months, however.

Remarks.—Resident, and partially migratory, larger numbers visiting us in winter than stay to breed in summer. Notes : *quack*, loud and high-

WILD DUCK.

sounding when uttered by the female, and harsh
and low when by the male. Local and other names:
Mallard, Stock Duck, Common Wild Duck. Sits
closely, and covers over her eggs when leaving her
nest voluntarily.

DUNLIN.

Description of Parent Birds.—Length about eight
inches. Beak rather long, nearly straight, and
black. Irides brown. Crown and upper parts
reddish-brown, streaked on the head and back of
neck with dusky-black, and each feather on the
back being black in the centre. Wing-coverts
greyish-brown edged with light grey; quills dusky-
black, inclining to brown on some of the lesser,
which are greyish-white on the edges of the outer
webs. Upper tail-coverts white, quills ashy-brown
edged with grey, excepting the two centre feathers,
which are longer than the rest and dusky-brown.
Chin white, cheeks, throat, and sides of neck and
breast whitish, streaked with dusky-black; belly
and under-parts white. Legs, toes, and claws
dusky-black, slightly tinted with green.

The female differs little from the male, but is,
as a rule, slightly larger.

Situation and Locality.—On the ground, well
hidden by a tuft of ling, heather, or tussock of
coarse grass, in boggy, marshy, tarn-besprinkled
parts of moors and heaths in the north and extreme west of England; also in Scotland and Ireland. Our illustrations are from photographs taken
on the Westmoreland hills, between Shunner Fell
and Nine Standards.

Materials.—A few straws or bents forming a

DUNLIN'S NEST.

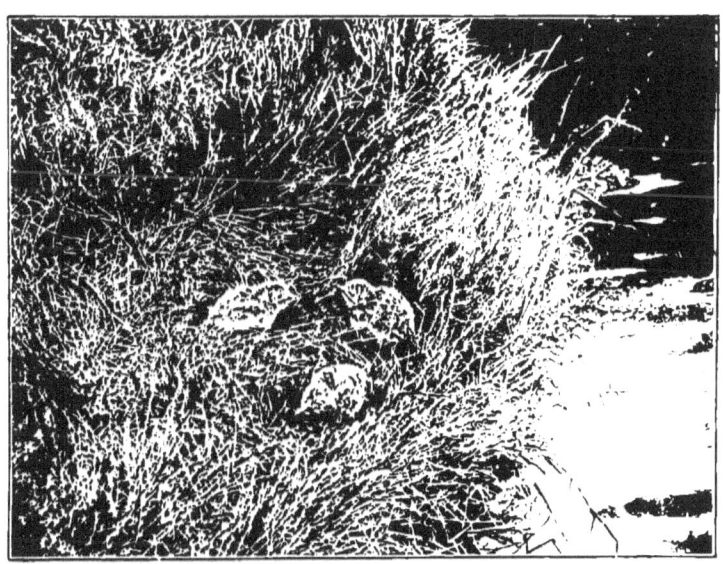

YOUNG DUNLINS.

slight lining to the hollow in which the eggs are deposited.

Eggs.—Four, pear-shaped; ground colour varies from greenish-white to cream or buff of different shades, blotched and spotted with reddish- and blackish-brown, and underlying markings of grey. Size about 1·3 by ·95 in.

Time.—May and June.

Remarks.—Resident, but migratory, partially and locally—that is to say, more birds visit our coasts in winter than stay to breed, and those that do breed with us resort to the coast-line in winter. Notes: call, *kwee-kwee, trui,* or *pe, pe, pe.* Local and other names: Dunlin Sandpiper, Purre, Judcock, Stint, Oxbird, Plover's Page, Churr, Sea Snipe, Sea Lark, Least Snipe. Sits pretty closely.

EAGLE, GOLDEN.

Description of Parent Birds.—Length about thirty-six inches. Beak moderately long, much curved at the tip, and bluish horn colour; bare skin round the base yellow. Irides hazel. The whole of the plumage is brown; the head, back of neck, and some of the wing-coverts reddish; wing-quills blackish-brown; tail-quills of two shades of brown, darkest at the tip. Chin and throat dark brown; under-parts of the body and thighs bay. The legs are feathered down to the feet, which characteristic distinguishes this bird from the Sea Eagle. The feet are yellow and the claws black. Mr. Booth was of opinion that the Golden Eagle does not assume the full mature plumage until it is five or six years old.

The female resembles the male in plumage, but is somewhat larger in size.

GOLDEN EAGLE'S EYRIE, WITH YOUNG, AND PARTLY DEVOURED PREY.

Situation and Locality.—On ledges of high inaccessible cliffs and precipices in the wildest and most desolate parts of Scotland and Ireland. In some of the Highland deer forests this noble bird is now strictly preserved, and such most commendable hospitality will no doubt save it to us for some time to come. Our illustration is from a photograph taken in the Western Isles of Scotland, but in the interests of British ornithology I think it best not to advertise the exact spot. The nest contained two partly-consumed mountain hares, off which the down had nearly all been carefully plucked, and the hind legs of a half-grown black rabbit.

Materials.—Sticks, bits of heather, dead fern-fronds, grass, and moss. The nest is repaired from year to year, and consequently often becomes a very bulky structure, on account of the bird using the same site for a long period. The subject of our illustration contained a large quantity of sticks and rubbish.

Eggs.—Two, sometimes three; very rarely four. Subject to variation both in ground colour and markings. The commonest type is dingy white, clouded, blotched, and spotted nearly all over with rusty or reddish-brown, and underlying markings of grey. Some specimens are pure white, unspotted. Size about 2·9 by 2·35 in.

Time.—March and April.

Remarks. — Resident but wandering. Note : " A barking cry " according to Seebohm ; and " a loud yelp uttered several times in succession " according to Dresser and Sharpe. Local and other names : Ring-Tailed Eagle, from the fact that young specimens have the basal half of the tail white. Sits closely.

EAGLE, WHITE-TAILED. *Also* Sea Eagle.

Description of Parent Birds.—Length about twenty-eight inches. Beak somewhat lengthened and nearly straight, except at the tip, where the upper mandible is much hooked. It is very strong, horn colour at the tip, and yellow at the base, as is also the bare skin surrounding that part. Irides very light yellow. Head and neck ash-brown, varying in hue with age ; back and wings dark brown, a few lighter-coloured feathers being intermingled ; wing-quills dusky-black. Tail white. Breast and under parts dark brown. Legs and toes yellow ; claws black.

The female is larger than the male, and both are subject to great variation in the colour of their plumage.

Situation and Locality.—On ledges and in holes of high inaccessible cliffs, generally near the sea ; in a tree or upon the ground on a small rocky island in the middle of a mountain loch. On the West Coast of Scotland and the surrounding islands, and in Ireland.

Materials. — Sticks, twigs, seaweed, heather, grass, and wool. The nest is often a huge structure, from the fact that the bird adds to it year by year. Specimens have been known measuring as much as five feet across. It is very shallow.

Eggs.—Two generally, sometimes only one ; and upon exceptional occasions three have been found. White, usually quite unspotted, but upon rare occasions specimens have been found slightly marked with pale red. Average size about 3·0 by 2·25 in.

Time.—March, April, and May.

Remarks.—Resident, but wandering. Note: a yelping, or barking kind of cry. Local and other names: Sea Eagle, Erne, Cinerous Eagle. Sits rather lightly, and is much attached to the same nesting site.

FALCON, PEREGRINE.

Description of Parent Birds.—Length from fifteen to seventeen or eighteen inches. Bill short, strong, much curved, and blue with a blackish tip. Bare skin round the base of the beak and eyelids yellow. Irides dusky. Head, back of neck, and upper parts generally bluish-ash, coloured darkest on the crown and nape, and faintly barred on the back and wing-coverts with a darker tint. Wing-quills dusky, barred and spotted on the inner webs with reddish-white. Tail-feathers barred alternately with black and dingy ash. Chin, throat, and upper breast white, tinged with yellow or rufous, and marked on the two latter parts with a few dark streaks. Lower breast, belly, and under-parts white, barred with dark brown and grey. Legs and toes yellow; claws black.

The female is somewhat larger, but similar in plumage. However, the species is subject to a great amount of individual variation.

Situation and Locality.—On ledges and in the crevices of rugged inaccessible sea cliffs and inland crags. In one or two places in England and Wales, and more numerously in Scotland and Ireland. I know a scaur in Westmoreland where the bird frequently attempts to breed, but invariably gets shot or robbed. The illustration appearing on p. 15 represents a cliff in Mull, in which a Peregrine Falcon and Common Buzzard were nesting at the time the photograph was taken. The former was in a horizontal crevice, out of which we managed to scare the female by the report of a gun. After chattering loudly overhead for some time, she alighted on a projecting crag far out of shot-reach above us, and afforded me the pleasure of examining her leisurely through my field-glasses.

Materials.—Sticks, dry seaweed, heather, and wool, or hair, bones, and castings; often no materials whatever.

Eggs.—Two to four. Morris says on rare occasions even five. Ground-colour varies from light orange-yellow to pale russet-red, thickly spotted, clouded, and mottled with reddish-brown of various shades. Size about 2·05 by 1·6 in.

Time.—April and May.

Remarks. — Resident. Note: a loud chatter. Local and other names: none. Sits lightly or closely, according to position, and is particularly partial to an old nesting site.

FLYCATCHER, PIED.

Description of Parent Birds. — Length about five inches. Bill rather short, straight, pointed, and black. Irides dark brown. On the forehead is a small white patch; crown and nape brownish-black; back black; wing-coverts and quills blackish-brown; edges of greater coverts and outer webs of tertials white. Tail dusky-black, parts of outer and second feathers white. All the under parts are white. Legs, toes, and claws black.

The female lacks the white forehead, and is generally less distinctive in her coloration.

Situation and Locality.—In holes of trees and walls and crevices of rock, in wild, out-of-the-way

G

parts in the six northern counties of England, Wales, and the South of Scotland. In the days of my youth I knew an old ruin in Yorkshire wherein a Pied Flycatcher and a Redstart nested within a few feet of each other. The former bird occupied the site for years in succession.

Materials.—Dry grass, moss, leaves, feathers, and hair, loosely put together.

Eggs.—Five to eight, generally five or six. Of a uniform pale blue or greenish-blue, closely resembling those of the Redstart, but are occasionally marked with a few reddish-brown spots, it is said; however, I have never seen any so marked. Size about ·75 by ·55 in.

Time.—May and June.

Remarks. — Migratory ; arriving in April and leaving in September or October. Notes very like those of the Redstart. Local and other names : Coldfinch. A close sitter : I could often have caught the bird whose nest I have just mentioned.

FLYCATCHER, SPOTTED.

Description of Parent Birds. -Length about six inches. Bill of medium length, straight, broad at the base, and dusky-black in colour. Irides dark brown. Head, back of neck, back, rump, and upper tail-coverts brown, the head being spotted with a darker tinge of the same colour. Wings brown, tail the same colour, very slightly forked, and a trifle lighter at the tip. Chin, throat, breast, and under parts a dull white, streaked on the throat and breast with dusky-brown. Legs, toes, and claws dusky-black.

The female is very similar to the male.

SPOTTED FLYCATCHER.

Situation and Locality. — On the horizontal branches of fruit trees trained against walls, in trellis-work, rose trees trained against houses, in holes in walls, ivy climbing up a wall or the trunk of a tree (as in our illustration), on ledges of rock, and in almost every conceivable situation.

Materials. — These vary as considerably as the positions selected for their accommodation. Straws, fibrous roots, moss, hair, feathers, rabbits' down, and cobwebs, somewhat loosely put together, as a rule, but occasionally I have come across a very compact little structure.

Eggs. — Four to six, generally five, varying considerably in coloration. The ground-colour ranges from grey to light green, the markings running through various shades of faint red or reddish-brown. Sometimes they are almost entirely absent, at others they form a belt round the larger end, and I have met with them with large, bright rust-red spots thickly distributed over the entire surface. Size about ·75 by ·57 in.

Time. — May, June, and July.

Remarks. — Migratory, arriving in the early part of May and leaving in September and October. Notes: a weak chirp and a harsh call-note. Local and other names: Beam-bird, Rafter, Bee-bird, Chanchider, Cherry-sucker, Bee-eater, Post-bird, Cherry-chopper. Sits closely, and flies away without demonstration when disturbed.

GADWALL.

Description of Parent Birds.—Length about twenty-one inches. Bill of medium length, broad, flat, and leaden-coloured. Irides hazel. Head and upper portion of neck pale brown, mottled with a darker tinge of the same colour; back grey, of two shades running in alternate curved lines; rump and upper tail-coverts bluish-black; wings long and pointed, small coverts reddish-brown, greater nearly black; secondaries brownish-grey, with a conspicuous white patch on them; primaries brown; tail-quills darkish brown, bordered with a lighter tinge of the same colour; lower half of neck dark grey, marked with short curving lines of a lighter tinge; breast and belly white; sides, flanks, and vent marked with irregular vertical lines of two shades of grey; lower tail-coverts black. Legs, toes, and webs dull orange; claws black.

The female has the head and upper part of the neck pale brown, spotted with dark brown; back of neck, back, and rump brown, the feathers being edged with pale reddish-brown; the wings are similar in markings to those of the male, but not so bright; lower part of neck, in front, and breast pale brown, with broad curved bands of dark brown.

Situation and Locality.—On the ground amongst reeds, sedges, rushes, and long, coarse grass on small islands situated in lakes; on the banks of broads, pools, in marshes and swamps in Norfolk only, so far as is known. Our illustration was obtained in that county.

Materials.—Dead grass, sedges, or leaves, lined with down of a brownish-grey colour, obscurely tipped with white. The tufts are smaller than those found in the Mallard's nest.

Eggs.—Five to thirteen, generally from eight to ten. Buffish-white and polished, closely resembling those of other members of the Duck family. Size about 2·1 by 1·5 in.

Time.—May and June.

Remarks.—At one time only a rare winter visitor, but now a resident with us. Some forty odd years ago a pair captured in a Norfolk decoy were pinioned and turned loose. They bred and multiplied and, it is thought, induced migrants to stay and do the same, until now there is a very respectable number in the county above mentioned. Notes: a low quacking. Local and other names: Gadwall Duck, Grey Duck, Common Gadwall, Rodge. Sits close.

GANNET.

Description of Parent Birds.—Length about thirty-four inches. Bill about six, straight, broad at the base, and horny greyish-white in colour. Irides pale straw colour. Skin of face and throat bare and blue. Head and neck buff. The whole of the body white, except wing primaries, which are black. The tail is tapering and pointed. Legs, toes, and webs black, with a pea-green line running up the front of each shank.

The female closely resembles the male. Some authorities say that the bird does not don its adult plumage until it is three years old; others place the age limit at four.

Situation and Locality.—On the shelves and ledges of precipitous sea cliffs and rocks. The birds breed in colonies, and engage every available situation capable of accommodating a nest. On

GADWALL.

the Bass Rock, Lundy Island, and many suitable places on the islands lying to the north and west of Scotland; also off the Irish coast. Our illustration was procured at Ailsa Craig.

Materials.—Seaweed, bits of turf, moss, and grass, sometimes in large quantities.

Egg.— One. White or bluish-white, covered like that of the Cormorant, with a thick coat of lime, which quickly becomes soiled and dirty by being trodden upon. The one represented in our illustration (forming the frontispiece to this work) had the thick end pointing directly towards the camera, so that the foreshortening has given it a somewhat round appearance. Size about 3 by 2 inches.

Time.—May and June.

Remarks.—Resident, but subject to much local movement. Notes loud and harsh. Local and other names: Solan Goose, Common Gannet, Solan Gannet, Soland Goose. Gregarious, and sits very close.

GARGANEY.

Description of Parent Birds. — Length about sixteen inches. Bill fairly long, straight, and black. Irides light hazel. Crown and back of head dark brown, which colour passes down the back of the neck, ending about the middle in a point. Back dark brown, the feathers being bordered with a lighter tinge of the same colour. Wing-coverts ashy-grey, scapulars elongated and narrow, white in the centre, and black round the edges; the transverse reflecting patch on the secondaries is green, bordered with white; primaries brownish-black; tertials grey; tail greyish-brown. A white stripe commences in front of the eye, passes over it and the ear-coverts, and becoming narrower, runs down the side of the neck for some distance. Cheeks and sides of neck reddish-brown, interspersed with fine lines of white pointing downwards. Chin black; throat and breast dark brown, marked with short, semicircular lines of light brown. Belly white; sides and flanks crossed with wavy black lines, which terminate towards the vent in two wide bands. Vent and under tail-coverts mottled with dusky-black. Legs, toes, and webs greyish-brown.

The female differs considerably from the male. She is smaller in size; her head is brown, marked with lines and spots of a darker tinge; back and wing feathers closest thereto dark brown, bordered with rusty brown, and tipped with white; wing-coverts greyish-brown, and green patch on wing duller. The white band over the eye is duller, and tinged with yellow. Chin white; breast greyish-white, marked with two shades of brown. Sides and flanks light brown, marked with a darker tinge of the same colour.

Situation and Locality.—On the ground in a tuft of rushes or sedge; amongst reed beds and coarse rank herbage on the rough banks of broads, rivers, and marshy pools in Norfolk and Suffolk, where alone the bird is now said to breed, and, pleasant to state, to be on the increase.

Materials.—Rushes, leaves, dry grass, and small brown tufts of down with long white tips from the bird's own body.

Eggs.—Eight to thirteen or fourteen. Creamy white, of varying shades, very similar indeed to those of the Teal, but perhaps a trifle more creamy in tint. Size about 1·8 by 1·35 in.

Time.—April and May.

Remarks.—Migratory; arriving in February and March, and departing in November. Note: a loud, harsh *knack*. The male has a peculiar, rattle-like note in the spring. Local and other names: Garganey Teal, Garganey Duck, Summer Duck, Summer Teal, Cricket Teal, Crick, Pied Wigeon. Sits very closely.

GOATSUCKER. *See* NIGHTJAR.

GOLD-CREST. *Also* GOLDEN-CRESTED WREN.

Description of Parent Birds.—Length about three and a half inches. Bill rather short, straight, slender, and black. Irides hazel. Forehead and round the eyes whitish, tinged with dull olive-green. Crown pale orange in front, and darker and richer towards the hind part. The feathers are somewhat elongated, and form a kind of crest, which is bounded on either side by a black streak. Neck, back, rump, and upper tail-coverts olive-green. Wing-quills dusky black, edged with greenish-yellow; coverts black, tipped with white, forming two white bars on wings, plainly visible during flight. Tail-quills dusky, edged with yellowish-green. All the under parts are greyish-white, tinged with buff on the throat, breast, and sides. Legs, toes, and claws brown.

The female is less distinct in coloration, and her crest is somewhat modified in size.

Situation and Locality.—Usually suspended from the branch or branches of a spruce fir; sometimes a cedar, yew, or holly is selected. It is placed near the end of a horizontal branch, at a height of from

GOLD-CREST.

two or three to ten or twelve feet from the ground; rarely in bushes; in woods, plantations, spinneys, shrubberies, and small clumps of trees, pretty generally throughout the United Kingdom where suitable trees are plentiful. Our illustration is from a photograph taken on the outskirts of a large plantation in Norfolk.

Materials. — Green moss, lichens, fine grass, spider-webs, caterpillar cocoons, and hair, beautifully felted together, and lined with down and feathers. It is a wonderfully compact little structure, for which its builder has been known to steal materials from the nest of a Chaffinch close by.

Eggs. — Four to ten; generally six or seven. Pale flesh colour, or very faint brown; occasionally white, spotted, and suffused, at the larger end generally, with light reddish-brown. Size about ·56 by ·42 in.

Time.—March, April, May, and June.

Remarks.—Resident, and a winter visitor. Notes: song weak but pleasant. Call: a shrill *tsit, tsit.* Local and other names: Golden-Crested Wren, Golden-Crowned Knight, Golden-Crested Warbler, Gold-Crested Wren, Gold-Crowned Wren, " Woodcock Pilot," from the fact that, as a winter visitor, it precedes that bird by a few days. A close sitter, and the smallest British bird.

GOLDEN EYE.

Is said to have bred in the North of Scotland, but no reliable ornithological authority has yet verified the statement, so far as I can gather.

GOLDFINCH.

Description of Parent Birds.—Length about five inches. Bill rather short, nearly conical, whitish at the base, and black at the tip. Forehead and chin rich scarlet, divided by a line of black, which passes from the base of the beak to the eyes. Cheeks white. Crown and back of head black, which descends on either side of the neck in a narrowing band. Back and rump pale tawny-brown, lightest on the back of the neck. Wing-coverts black; quills black, barred across with yellow, and tipped with white. Upper tail-coverts grey, mixed with tawny-brown; quills black, marked with white and buffy-white spots near their tips. Throat and under parts white, tinged on the breast, sides, and flanks with pale tawny-brown. Legs and toes pale pinkish-white; claws brown.

The female is somewhat similar, but is said to have a more slender beak, the red on her head to be less in area, and often speckled with black, and the smaller coverts of the wings to be dusky-brown instead of black.

Situation and Locality.—In the fork of an apple, pear, or other fruit-tree in gardens and orchards; on the boughs of chestnut and sycamore trees, evergreens, and sometimes in thick hedgerows. Sparingly throughout England, in some parts of Scotland, and widely, though not numerously, in Ireland.

Materials.—Moss, fine roots, dry grass straws, bits of wool, lichens, and spiders' webs, lined with feathers, willow down, and hairs. It is a neat little cup-shaped structure, the materials of which depend to some extent upon what the bird may find lying around.

Eggs.—Four to five, greyish, or greenish-white, spotted and streaked with light purplish and reddish-brown, and grey. The markings are, as a rule, most numerous round the larger end of the egg. They generally run smaller in size than those of the Greenfinch and Linnet, but closely resemble them in colour. Size about ·66 by ·51 in.

Time.—May, June, and July.

Remarks.—Migratory and resident. Notes: call, *ziflit*, or *sticklit*. Song: shrill twittering and warbling, and containing the syllable *fink*. Local and other names: Gold Spink, Draw-water, Thistle Finch, Grey Kate, or Pate, Goldie, King Harry, Redcap, Proud Tail. Sits pretty close, and flies away without demonstration.

GOOSANDER.

Description of Parent Birds.—Length twenty-six and a half inches. Bill rather long, straight, hooked at the tip, and vermilion-red, except the upper ridge and point of the upper mandible, which are black. Irides red. Head and upper half of neck rich glossy green; feathers on the back of the head lengthened. Upper back and scapulars black; lower back, upper tail-coverts, and tail-quills ash grey. Shoulder of wing, all the coverts and secondaries white; primaries almost black. Lower part of neck in front, breast, belly, vent, and under tail-coverts salmon-buff. Legs and toes orange-red, webs somewhat darker.

The female is rather smaller, and differs to a considerable extent in coloration. Bill and irides duller. Head and upper part of neck reddish-brown. Back, wings, tail-coverts, tail-quills, sides and flanks ash-grey, except secondaries and primaries, which are white and lead-grey respectively. Throat white, breast and under parts tinted with buff. Legs and feet orange-red.

Situation and Locality.— Holes in trees, clefts in rocks, holes amongst the exposed roots of trees, on ledges of rock, under the cover of bushes, on small islands, in fresh-water lochs, on the banks of streams and lochs, in forests in the northern Highlands. No absolute proof of the bird's nesting in the British Isles was forthcoming until as late as 1871.

Materials.—Depend somewhat upon position; those in trees are said to have none except the decayed wood, and down from the bird's own body, whilst in other situations liberal quantities of dead weeds, dry grass, and small roots and down are said to be used.

Eggs.—Six to twelve. Creamy-white. Size about 2·7 by 1·85 in.

Time.—April and May.

Remarks.—Migratory, being principally a winter visitor, but a few remaining to breed. Note: a low plaintive whistle. Local and other names: Dun Diver, Saw Bill, Jacksaw, Sparling Fowl. Sits close.

GOOSE, GREY-LAG.

Description of Parent Birds.—Length thirty-five inches. Bill of medium length, fairly straight, and pink flesh-colour, except on the tip of each mandible, where it is white; irides brown; head, back of neck, and upper portion of back, ash-brown: the feathers of the last bordered with a lighter tinge; wings lead-grey on the portions nearest the back, each feather being broadly margined with lighter grey, the outer front portion pale bluish-grey, the rest dark leaden-grey; lower portion of the back and rump light bluish-grey; upper tail-coverts white; tail-quills white on the inside webs and greyish-brown on the outer, and tipped with white; chin, throat, and breast light grey; belly, vent, under tail-coverts, and under side of tail-quills, white. Sides, flanks, and thighs barred with two shades of grey; legs, toes, and webs flesh colour; claws black.

The female is smaller in size than the male.

Situation and Locality.—On the ground amongst tall rank grass, heather, rushes or osiers in lonely swamps and moorland bogs of Ross, Sutherland, Caithness, and the outer and inner Hebrides; also in a semi-domesticated state at Castle Coole, in Ireland, where the birds are only subject to a limited local movement. Our illustration was procured on a small island in a fresh-water loch in the Outer Hebrides.

Materials.—Heather, dried flags, rushes, leaves, and grass in large quantities, with an inner lining of feathers and down from the breast of the female.

Eggs.—Five to nine, sometimes twelve to fourteen; but all the best authorities have had to accept the latter numbers on hearsay. Dull yellowish, or creamy-white, with a very slight suggestion of green. Morris says they are smooth and shining, but Mr. Dixon and Messrs. Sharpe and Dresser say they are dull and without polish; and the statements of the latter authorities are, of course, borne out by specimens in collections. Size about 3·4 by 2·35 in.

Time.—March, April, and May.

Remarks.—Resident, but subject to much local

YOUNG GREY-LAG GEESE.

II

movement, and numbers increased by northern
arrivals during winter. Note: a "gaggle." Local
and other names : Wild Goose, Grey Goose, Grey-
legged Goose. Sits close.

GOSHAWK

Has now quite ceased to breed within the
British Isles, and is only a straggler seen upon
rare occasions.

GREBE, GREAT CRESTED.

Description of Parent Birds.—Length about
twenty-two inches. Bill rather long, straight,
pointed, black at the tip, and reddish towards the
base. The top of the head and the divided crest
with which it is adorned are dusky ; cheeks whitish.
Round the upper part of the neck is a tippet or
ruff, which is formed of elongated feathers that
stand out all round. These feathers are rusty red,
with a darker tinge at the tip of each. Hind part
of the neck, back, wings, and short, tufty tail, dark
brown, except the secondaries of the wings, which
are white. Front of neck, breast, and belly, silvery
white. Sides and flanks, pale chestnut ; outside of
legs and toes, dusky green; inside, pale yellowish-
green. Each toe is surrounded by a margin of
web.

The female is not so large or distinct in
coloration. Her crest is also smaller.

Situation and Locality.—Amongst reeds growing in the water. Sometimes its foundation rests
upon the bottom, at others it is moored to the

surrounding vegetation. On large sheets of fresh water. It breeds on the Norfolk and Suffolk Broads, in Wales, Yorkshire, Shropshire, Cheshire, Lancashire, and several other counties. Mr. Saunders and other naturalists say that it does not breed in Scotland; but Mr. Dixon, in whom I place great reliance, says that it does so in the southern counties; however, I am unable from my own observations to confirm this. It is also found breeding in several parts of Ireland.

Materials.—Flags, sedge leaves, reeds, and all kinds of dead water-plants heaped together. The nest has a slight hollow on the top, and does not stand far above the level of the water.

Eggs.—Three to five, usually four. White when originally laid, but soon becoming stained and dirtied. Size about 2·2 by 1·45 in.

Time.—April, May, and June.

Remarks.—Resident but wandering. Note: a harsh, single-syllabled kind of croak. Local and other names: Gaunt, Molrooken, Loon, Tippet Grebe, Greater Loon, Cargoose. Gregarious. Covers over eggs on leaving nest voluntarily. Makes several mock nests, supposed to be either for the male, as outlook posts, or for the young ones when hatched. Sits lightly, and dives when the nest is approached.

GREBE, LITTLE. *Also* DABCHICK.

Description of Parent Birds.—Length about ten inches. Bill not very long, straight, and brown. Irides reddish-brown. Crown, back of neck, and the whole of the upper parts, dark rusty-brown. Cheeks, throat, and sides of neck, reddish-brown; breast, belly, and under parts, greyish-white. Legs and toes, dark greenish. The wings are short, tail almost nil, and legs situated far behind.

The female is very similar to the male.

Situation and Locality.—Amongst reeds, rushes, weeds, and long, coarse grass growing on or near the banks of pools, sluggish rivers, lakes, reservoirs, and canals. The nest is a floating kind of raft, built up from the bottom in all suitable localities throughout the British Isles.

Materials.—A liberal collection of dead, half-rotten, aquatic weeds, thoroughly saturated with water; very shallow at the top.

Eggs.—Four to six; as many as seven have upon a few occasions been found. White, and rough-surfaced when first laid, but gradually becoming stained and discoloured by contact with the bird and the decaying weeds upon which they are deposited, and are often covered by. Size about 1·45 by 1·0 in.

Time.—March, April, May, June, July, and August.

Remarks.—Resident, but subject to local move-ment. Note: alarm, *whit, whit.* Local and other names: Dabchick, Black-chin Grebe, Small Ducker, Didapper, Dobchick, Loon, Dipper. Not a close sitter, but covers over its eggs when leaving the nest voluntarily.

GREENFINCH.

Description of Parent Birds.—Length about six inches. Bill, short, thick, and flesh-coloured. Irides hazel. Head, neck, back, rump, and upper tail-coverts, yellowish-green, mixed with ashy-grey on the sides of the head and neck, and with greyish-

brown on the other parts. The forehead and rump
are bright golden-green. Wing-quills dusky, some
of them bordered with yellow and others with
grey on the outer webs. Tail feathers dusky,
those in the middle uniform, the rest bordered
with yellow on their exterior webs. Chin, throat,
and breast, bright yellowish-green; belly lighter
and mixed with ash-grey; vent and under-tail-
coverts white, tinged with pale yellow. Legs, toes,
and claws light pinkish-brown.

The female is somewhat smaller, and her upper
parts are greenish-brown, tinged only with yellow
on the wing-coverts, rump, and wing and tail quills;
but this is of a duller character than that found on
the feathers of the male. Under parts dull greyish-
brown, inclining to greenish-yellow on the belly.

Situation and Locality.—In thick, whitethorn
hedges, gorse bushes, yew-trees, ivy, holly, and
other evergreens; in shrubberies, orchards, on
commons, and almost anywhere in suitably-wooded
districts. Our illustration is from a photograph of
one out of four, situated within a few yards of each
other in a thick hedge dividing an orchard from a
Surrey common. It is met with in suitable localities
throughout the United Kingdom.

Materials.—Slender twigs, rootlets, moss, and
grass, lined internally with hair and feathers.

Eggs.—Four to six, white, pale grey, or white
tinged with blue, in ground colour, sparingly spotted
with varying shades of brown, from greyish to
dark liver-coloured. The spots and markings are
generally most numerous at the larger end.
Specimens have sometimes been found pure white
and unmarked. They are often very difficult to
distinguish from the eggs of the Goldfinch. Size
about ·82 by ·56 in.

Time.—April, May, June, July, and sometimes as late even as August.

Remarks.—Resident. Notes: call when flying, *yack-yack*, and when perched, *shwoing*, according to Bechstein. When disturbed whilst sitting, it utters a heart-softening sort of melancholy *tway* that is enough to fill any young collector with remorse. Local and other names: Green Linnet, Green Chub, Green Grosbeak, Green Bird, Green Lintie. Sits very closely.

GREENSHANK.

Description of Parent Birds.—Length about twelve or thirteen inches. Bill long, slightly curved upwards, and almost black in colour. Irides hazel. Head, sides, and back of neck greyish-white, marked with almost black longitudinal lines. Back and wings (except primaries, which are dull black) greenish-black, each feather being bordered with buffy-white. Tail-quills white, barred in the middle and striped on the outside with ash-brown. Chin, throat, breast, sides, belly, vent, and under tail-coverts white, the throat and sides being slightly streaked with ash-grey. Legs and toes olive-green ; toes black.

Female similar to male.

Situation and Locality.—On the ground amongst tufts of coarse grass, heather, between dry mounds near lochs and streams in the North and West of Scotland, the Hebrides, and Shetlands.

Materials.—A few bits of dead grass, used as a lining to the declivity chosen.

Eggs.—Four ; pale yellowish-green to warm stone-colour or buff, beautifully blotched or spotted with light purple, grey, and dark brown. Markings

GREENFINCH.

most numerous at larger end. Size about 1·95 by
1·35 in.

Time.—May.

Remarks.—Migratory, arriving on its breeding-
grounds at the end of April or beginning of May,
and leaving in July. The bird is said to winter
in Ireland. Note: a loud *chee-weet, chee-weet.*
Local and other names: Cinerous Godwit, Green-
legged Horseman, Greater Plover. Sits close, and
is very demonstrative when disturbed.

GROUSE, BLACK.

Description of Parent Birds. — Length about
twenty-two inches. Bill short, curved downwards,
and black. Irides dark brown. Bare, erectile skin
over eyes, bright scarlet. Head, neck, back, wing-
coverts, rump, and tail black, richly glossed in
parts with blue-black. Wings brownish-black, with
a conspicuous white bar across the middle. The
tail feathers are elongated on either side, and form
an outward kind of curving hook. Chin, breast,
belly, and flanks black; vent, thighs, and legs dark-
brown, mixed with white; under tail-coverts white;
toes and claws blackish-brown.

The female is shorter by four or five inches,
and differs considerably in appearance. Her bill
is dark brown. Irides hazel. Plumage red or
rusty-brown, barred and freckled with black; the
markings are largest on the breast, where the
feathers are bordered with greyish-white. The tail
is not forked, and the feathers are variegated with
rusty-red and black, and tipped with white. Under
tail-coverts nearly white. Legs mottled brown;
toes and claws brown.

BLACK GROUSE.

Situation and Locality.—On the ground, under tufts of dead bracken, brambles, heather, rushes, and coarse grass. I have seen them quite exposed in open pasture land, and have known cows tread upon and break their eggs in such situations. On rough broken land containing heather, rushes, ling, gorse, juniper, mixed woods and young plantations. The bird breeds in suitable parts of England, Wales, and Scotland, but not in Ireland. Our illustration is from a photograph taken in the Highlands of Scotland.

Materials.—Dry grass, bents, fern or bracken fronds, and other suitable materials at hand, forming a scant lining to the selected hollow.

Eggs.—Five to ten ; yellowish-white to yellowish-brown, irregularly spotted with smallish red-brown spots. Size about 2·0 by 1·4 in. Distinguished from those of the Capercaillie by their smaller size.

Time.—April, May, and June.

Remarks.—Resident. Notes : male, a loud cooing, followed by a hissing sound ; female's response plaintive. Local and other names : Black Game, Heath Cock, Black Cock, Heath Poult, Grey Hen (female), Brown Hen (female). Sits closely.

—

GROUSE, RED.

Description of Parent Birds.—Length about sixteen inches. Beak short, curved downward, and black. Irides hazel. Above the eye is a scarlet, arched membrane. The dominating colour of the head, neck, back, wing, and tail-coverts is reddish-brown, speckled and barred with black. Wing and tail-quills blackish-brown. Chin and throat rich dark chestnut-brown, unspotted ; breast dark reddish-

brown, sometimes almost black; belly, sides, vent,
and under tail-coverts light reddish-brown, tipped
with white. Legs and toes covered with short soft
feathers of greyish-white; claws long and horn
colour, dark at the base.

The female is somewhat smaller and of a lighter
rufous-brown. The membrane over her eye is
narrower and less conspicuous. Both male and
female are subject to considerable variation in
plumage. I have seen some hens a beautiful golden
yellow, and both sexes often showing a considerable amount of white in their plumage.

Situation and Locality.—A slight hollow or
natural depression, generally well hidden by heather
or ling, occasionally amongst rushes or long coarse
grass on wild moors in Wales, Derbyshire, the six
northern counties of England, in every county of
Scotland, excepting perhaps one, and in suitable
parts of Ireland. Our illustration is from a photograph taken in Westmoreland.

Materials.—A few heather or ling shoots, or
bits of bent grass.

Eggs.—Five to nine; as many as thirteen to
fifteen have been found. Of a dirty white ground-
colour, thickly blotched and spotted with umber-
brown. Variable in regard to colour and markings,
but distinguished from those of the Ptarmigan by
being less buff in ground colour and more spotted.
Size about 1·75 by 1·25 in.

Time.—Eggs have been found as early as
February and as late as July; but April, May,
and June are the principal breeding months.

Remarks.—Resident. Notes: female call, *gor*,
gor, *gor*, pronounced with a peculiar nasal catch;
crow notes, *cabour*, *cabour*, *cabeck*, *cabeck*, *beck*,
beck; *cockaway*, *cockaway*; alarm note of male.

cock, cock, cock. Local and other names: Gorcock, Moorfowl, Moorcock, Moorgame. A close sitter, resorting to decoy methods when disturbed.

GROUSE, SAND.

Although several great invasions of this species have from time to time taken place, and the bird has during its latest, in 1888, bred with us, it has no rightful claim for inclusion at length in this work.

GUILLEMOT, COMMON.

Description of Parent Bird.—Length about eighteen inches. Bill rather long, straight, sharp-pointed (which easily distinguishes the bird from the Razorbill), and black. Irides dusky. Head, neck, back, wings (except ends of secondaries, which are tipped with white), and tail dark mouse-brown. Lower part of throat, breast, and belly white. Legs and feet, which are webbed, brownish-black.

The female is rather smaller than the male.

Situation and Locality.—On ledges and in hollows of cliffs, on the flat, bare summits of rock-stacks in suitable places pretty generally round our coasts. Our illustration shows a great number of these birds sitting on their eggs at the Farne Islands. The rock was so thickly covered with them that whenever a new-comer flew up out of the sea it was obliged to drop in the middle and, so to speak, elbow its way in. The birds kept up an incessant chattering, and how they ever recognise their own eggs I can't imagine.

RED GROUSE

Materials.—None whatever, the egg being laid on the bare rock.

Egg.—One, very large for the size of the bird, and pear-shaped. The eggs of this species present an endless variety of coloration. Sometimes the ground-colour is white, at others cream, yellowish-green, reddish-brown, pea-green-blue, purplish-brown, and every variety of shade between these colours, spotted, blotched, and streaked profusely with black, dusky brown, greyish-brown, and other tints in great variety. Some specimens are without spots, and I have seen others on Ailsa Craig and else-where closely resembling those of the Razorbill, but always more pyriform. Size about 3·25 by 1.95 in.

Time.—May and June.

Remarks.—Resident. Notes (of young) : *willock, willock.* Local and other names : Foolish Guillemot, Willock, Tinkershere, Scout, Tarrock, Lavy, Murre, Sea Hen, Marrock. A close sitter.

GUILLEMOT, BLACK.

Description of Parent Birds. — Length about fourteen inches. Bill fairly long, straight, and black. Irides brown. The whole of the plumage is black, with exception of a large patch on the coverts of each wing, which is white. Legs, toes, and webs vermilion-red; claws black.

The female is similar to the male in size and coloration.

Situation and Locality.—In deep crevices of rocks overhanging the sea; amongst large stones heaped loosely together; and occasionally under or between crags and large fragments of rock near the beach. Principally on the western and northern

COMMON GUILLEMOTS BREEDING ON THE PINNACLES
AT THE FARNE ISLANDS.

coasts of Scotland and the islands round about it, in a few suitable places round the coast of Ireland, and to a limited extent in the Isle of Man.

Materials.—None ; the eggs being laid on the bare rock or ground.

Eggs.—Two, white, faintly tinged with green, blue, or creamy-buff, spotted and blotched with ash-grey, reddish or chestnut brown, and very dark brown. Size about 2·35 by 1·6 in.

Time.—May and June.

Remarks.—Resident, but a southern wanderer in winter. Notes : a plaintive whine. Local and other names : Sea Turtle, Greenland Dove, Dovekie, Scraber, Tyste, Puffinet. Gregarious. Sits closely. Keeps to the open sea, except during the breeding season and when driven ashore by stress of weather.

GULL, BLACK-HEADED.

Description of Parent Birds.—Length about sixteen inches. Bill moderately long, nearly straight, and lake-red. Irides hazel ; eyelids crimson. Head and upper part of throat dark brown. Back and sides of neck white. Back and wings (except some of the primaries, which are black at the tips, and on some of the margins with white shafts), uniform lavender-grey. Tail-coverts and quills white. Lower front of neck, breast, and all under parts, white. Legs and feet lake-red ; claws black.

The female is similar to the male.

The above description is of a solitary pair shot whilst nesting in June on a northern moorland tarn. The Black-headed Gull is subject to considerable variation in plumage, not only in regard to season but age.

BLACK-HEADED GULL.

Situation and Locality.—On the ground, in a tussock of coarse grass, tuft of rushes, or a slight hollow on the bare ground; in swamps and bogs, at the edges of and on islands in tarns and lakes. In large colonies at a great number of suitable places throughout the British Isles. Two famous places in England are Scoulton Mere in Norfolk, where the bird has nested in thousands for upwards of three hundred years in succession, and at Pallinsburn in Northumberland. Although gregarious, I have frequently met with solitary pairs nesting on small mountain tarns. Our illustration is from a photograph taken late in the season at Scoulton Mere.

Materials.—Sedges, rushes, tops of reeds, and withered grass; generally in small quantities, sometimes quite absent, and at others in fairly large quantities, much depending upon the site chosen.

Eggs.—Two to three; usually the second number, and occasionally four, varying from pale olive-green to light umber-brown in ground-colour, blotched, spotted, and streaked with blackish-brown and dark grey. Size about 2·2 by 1·45 in. They are subject to great variation in regard to size, shape, and colour; but their small size generally and the presence of the parent birds easily distinguish them.

Time. — April, May, and sometimes as late as June.

Remarks.—Resident, but subject to much local movement. Notes: a hoarse cackle, resembling a laugh when quickly repeated. Local and other names: Red-legged Gull, Laughing Gull, Peewit Gull, Blackcap, Sea Crow, Hooded Mew, Brown-headed Gull, Mire Crow, Croker, Pickmire. Sits lightly, and clamours noisily overhead when disturbed. Gregarious, as a rule.

GULL, COMMON.

Description of Parent Birds. — Length about eighteen inches. Bill rather short, slightly curved downward at the tip, and yellow in colour. Irides orange-brown. Head and neck snowy-white. Back and wings French grey ; tips of wings black, spotted with white, on account of some of the primaries having white ends. Tail-coverts and quills snowy-white. Chin, throat, breast, belly, and vent snowy-white. Legs, toes, and webs greenish-yellow.

The female is similar in plumage, but slightly smaller in size.

Situation and Locality.—On the ground amongst heather and coarse grass ; on ledges and in crevices of rock round the coast of Scotland ; on islands ; in inland lochs and tarns ; also in suitable places in Ireland, but now nowhere in either England or Wales. Our illustration was procured on the West Coast of Scotland.

Materials. — Heather, dry seaweed, and dead grass. It may be observed that a somewhat large nest is built as a rule.

Eggs.—Two to four ; generally three, buffish-brown or dark olive-brown in ground-colour, spotted, blotched, and streaked with grey, dark brown, and black, irregularly distributed. Size about 2·25 by 1·65 in. The smallness of the spots and the size of the eggs enable collectors easily to identify them.

Time.—May and June.

Remarks.—Resident, but subject to local movement. Notes : a kind of squeal. Local and other names : Winter Mew, Sea Mew, Sea Mall or Maw, Sea Gull, Sea Cob, Blue Maa. Gregarious. A light sitter, and clamorous when disturbed.

GULL, GREAT BLACK-BACKED.

Description of Parent Birds.—Length about thirty inches. Bill of medium length, large and powerful; pale yellow, excepting a portion of the under mandible, which is orange; upper mandible turned down at the tip. Irides straw-yellow. Head and neck all round snowy-white. Back and wings black, with exception of the tips of the quills, which are white. Upper tail-coverts and tail-quills white. Breast and all under parts pure white. Legs, toes, and webs pale flesh-colour.

The female is similar, but somewhat smaller.

Situation and Locality.—On the ledges of maritime cliffs, on the tops of rocky islets in the sea and fresh-water lakes, known as rock-stacks; also on the ground, in marshes and moors; on the coasts of Dorset, Cornwall, Scilly, and Lundy; on the Welsh coast, but most abundant on the western and northern shores of Scotland and the islands lying round about; also in Ireland. Our illustration was obtained near the summit of a small rock-stack in a Highland sea-loch.

Materials.—Seaweed, heather, wool, and dry grass in variable quantities. Sometimes they are almost entirely absent.

Eggs. — Two to three, generally the latter number. Yellowish-brown or stone-colour to light olive-brown, blotched with slate-grey and dark brown. The spots are not very large, and generally distributed over the surface of the egg. Size about 3·1 by 2·1 in. The large size of the eggs and the small spots are distinguishing characteristics.

Time.—May and June.

Remarks.— Resident, but wandering during the

COMMON GULL.

non-breeding months. Note: a harsh croak or
laugh. Local and other names: Cob, Blackback,
Great Black and White Gull. Gregarious in some
parts and solitary in others. Not a close sitter,
but demonstrative when intruded upon.

GULL, HERRING.

Description of Parent Birds. — Length about
twenty-four inches. Bill of medium length, hooked
at the tip, and yellow, with an orange spot on the
lower mandible. Irides pale yellow. Head and
neck white. Back and part of wings light grey,
which distinguishes the bird from the Lesser Black-
backed Gull; quills blackish, tipped with white.
Breast, belly, vent, upper tail-coverts and tail-quills
pure white. Legs and feet flesh-colour. Variable
with age.

The female is often much smaller than the male,
but is similar in the coloration of her plumage.

Situation and Locality.—Ledges of sea cliffs,
low rocky islands, sometimes in marshes, such as
Foulshaw Moss in Westmoreland. Our illustrations
are from photographs taken on the Farne Islands,
where the bird breeds in company with the Lesser
Black-backed Gull. The eggs represented in our
illustration are those of the bird sitting on her nest
in the centre of the group of gulls. It breeds in
all suitable localities round the coasts of England,
Wales, Scotland, and Ireland.

Materials.—Seaweed and turf, lined with grass,
sometimes in liberal quantities, at others very scant,
or absent altogether. The grass used appears to
have often been obtained quite green.

Eggs.—Two to three, varying in ground-colour

GREAT BLACK-BACKED GULL.

from olive-green to buffish-brown, spotted and
blotched with dark brown and grey. Variable, and
bearing a very close resemblance to the eggs of
the Lesser Black-backed Gull, but said to run
larger by some authorities, and to be more spotted
and less blotched. The fishermen reckon to distinguish them from those of the above-mentioned
species by the paler ground-colour of the shell;
but the only sure method of identification I have
found is to watch the parent bird on to her nest.
Average size about 2·85 by 2·0 in.

Time.—May and June.

Remarks.—Resident. Call-notes, *hau-hau-hau*;
alarm, *ky-cok*. Local and other names: none. Sits
lightly, and watches the intruder closely from a
distance of eighty or a hundred yards. A terrible
egg-stealer.

GULL, KITTIWAKE. *See* KITTIWAKE.

GULL, LESSER BLACK-BACKED.

Description of Parent Birds.—Length about
twenty-three inches. Beak of medium length,
nearly straight, and yellow, with the exception of
an orange spot on the under mandible. Irides
straw colour. Head and neck all round pure white.
Back and wings dark slate-grey, some of the quills
being slightly tipped with white. Upper tail-
coverts and tail-quills white. Breast, belly, vent,
and under tail-coverts pure white.

The female is said to be a little smaller, and
the feathers of the back and wings to vary much
in tint with age and locality.

Situation and Locality.—On the ground, in

HERRING GULL.

LESSER BLACK-BACKED AND HERRING GULLS AT THE
FARNE ISLANDS.

hollows scooped out of the soft turf, on grass growing in nooks and on ledges of rock, on bare rocks, and on masses of dry seaweed. Our illustration is from a photograph taken on the Farne Islands, where a large colony breeds. On low rocky islands, ledges of cliffs, on islands in inland lakes, and in moss-bogs. On nearly all suitable places round our coasts, except the greater parts of the east and southern coasts of England.

Materials.—Seaweed, often in large quantities; grass, which appears to have been collected quite green; sometimes no materials whatever, the eggs being laid on the grass in a hollow.

Eggs.—Two to four, generally three. Very variable, from light drab to dark olive-brown; sometimes pale bluish-green, spotted, blotched, and streaked with ash-grey, pale brown, and dark liver-brown. Size about 2·6 by 1·85 in.

Time.—May and June.

Remarks.—Resident, but subject to much local movement. Notes: call, *ha, ha, ha,* or *an, an, an;* note of anger, *kyeok.* Local and other names: Yellow-legged Gull, Less Black-backer Gull. Not a very close sitter, but noisy and clamorous when disturbed. Gregarious.

HARRIER, ASH-COLOURED. *See* HARRIER, MONTAGU'S.

HARRIER HEN.

Description of Parent Birds.—Length about eighteen inches. Beak short, much curved, bluish-black, and surrounded at the base with black, bristly feathers. Bare skin immediately round base of

LESSER BLACK-BACKED GULL.

beak, and irides yellow. Head, neck, back, wings, and upper side of tail bluish or ash grey, except the wing-primaries, which are almost black. Some specimens have a mottled, rusty-brown spot on the nape. Chin, throat, breast, and belly bluish-grey, much lighter on the latter parts. Thighs, vent, and under tail-coverts white. Under side of tail-quills very light grey, faintly barred with a darker tinge. Legs and toes yellow; claws black.

The female measures about four inches longer; her bill is nearly black, and the bare skin round the base tinged with green. Irides reddish-brown. Crown and back of neck dark brown; round the face is a kind of ruff, the feathers of which are a mixture of brown and white. Back and wings umber-brown, except some of the coverts, which are edged with rufous, and the primaries, which are of a dusky colour. The tail-quills are dark brown tipped with rusty-red; the centre ones uniform in colour, and those on the sides barred with lighter rusty-brown. Throat and all the under parts reddish-buff, with a darker patch in the centre of each feather. Tail-feathers underneath barred with brownish-black and grey.

Situation and Locality.—On the ground, amongst tall heather, furze, and other bushes; on moors, commons, fens, and on wild, lonely mountain-sides. Its destructive habits amongst game birds have made the gamekeeper an especial enemy, and he has waged incessant war upon it for so long that it is now almost exterminated in England. It is said to breed in Cornwall, Devon, Somerset and one or two other western counties, Wales, and the North of England occasionally. Its nest occurs most frequently in the Hebrides, Orkneys, and

Highlands of Scotland; also in suitable parts of Ireland.

Materials.—Small sticks, sprigs of heather, and coarse grass; in sparing quantities where the nest is placed in a dry situation; but when a low, damp place is chosen, sticks, reeds, sedge, and flags are said by some observers to be used in liberal quantities.

Eggs.—Four to five, occasionally six. White, faintly tinged with blue or bluish-green; on rare occasions slightly marked with light rusty-red or yellowish-brown. They vary in size, and closely resemble those of the Marsh and Montagu Harriers. Size about 1·75 by 1·45 in.

Time.—May and June.

Remarks.—Formerly resident, now probably only migratory. It arrives in April or May, and departs in September and October. Notes : tremulous and Kestrel-like. Local and other names : male, Dove Hawk, Blue Hawk, or Miller; female, Ringtail; and in the Hebrides a Gaelic name signifying Mouse Hawk. The sexual difference in plumage was the cause of the birds being believed at one time to represent different species. Not a close sitter.

HARRIER, MARSH.

Description of Parent Birds.—Length about twenty-one inches. Beak short, curved, and bluish-black. Bare skin round the base of the beak, and irides yellow. Crown, sides of head, and nape pale rusty yellowish-white, streaked with darkish brown. Back dark brown, tinged with red, the feathers being bordered with a lighter shade. Wing-coverts and tertials varying, according to age, from dark reddish-brown to ash-grey; secondaries ash-grey; primaries varying from brownish-black to slate-grey. Tail ash-grey. Chin and throat almost white; breast and under parts reddish-brown, streaked with dark brown. Legs and toes yellow; claws black.

The female is larger, and slightly duller in her plumage. Both are subject to variation in colour, according to age.

Situation and Locality.—On the ground, amongst sedges, reeds, ferns, and under furze and other small bushes; rarely in trees. On low, marshy, reed- and water-covered land; also unfrequented moors. Professor Newton, in the latest edition of Yarrell, issued 1874, says that "the bird breeds regularly in Devonshire, Norfolk, and Aberdeenshire;" and Mr. Dixon, in his "Nests and Eggs of British Birds," issued just twenty years after, says that Norfolk is the only county in Great Britain in which the bird regularly attempts to breed. This is one among many of the facts which serve to illustrate the rapidity with which our rarer birds are being banished.

Materials.—Sticks, twigs, rushes, and reeds in rather large quantities, lined with dead grass.

Eggs.—Three to five or six. White, sometimes slightly tinged with pale bluish-green or milk-blue, and upon rare occasions marked with a few spots of rusty-red. Size about 1·95 by 1·55 in.

Time.—May.

Remarks. — Resident, but wandering. Notes : male, *koi* or *kai;* female, *pitz pitz, peep peep.* Local and other names : Duck Hawk, White-headed Harpy, Moor Harrier, Moor Buzzard, Puttock, Marsh Hawk, Bald Buzzard. Sits lightly.

HARRIER, MONTAGU'S. *Also* Ash-Coloured
Harrier.

Description of Parent Birds.—Length about seventeen inches. Beak short, upper mandible much curved and nearly black. Skin round base of beak bare, and greenish-yellow. Irides bright yellow. Head, neck, back, and wing-coverts bluish-grey. Primaries nearly black; secondaries marked by three bars. Tail-quills, on the sides, white, barred with bright rust colour; centre feathers bluish-grey. Chin and throat brownish-grey; breast, belly, and under parts white, streaked with bright rust colour. Legs and toes yellow; claws black. The wings are very long and narrow.

The female is about nineteen inches long. Beak black; bare skin at base, dull yellow. Irides hazel. Crown and back of head reddish-brown, with spots of a darker tinge. Over and under the eye is a streak of grey. Back and wings dark umber-brown; rump and upper tail-coverts orange-brown and white. Side feathers of tail barred with brown of two shades; breast and all under parts light reddish-brown; claws black. Both sexes of this bird are said to vary considerably, according to age and individual.

Situation and Locality.—On the ground, amongst heather, ferns, long grass or rushes, furze, and low brushwood; on moors and heaths in Norfolk, Kent, Pembrokeshire, Dorsetshire, Hampshire, Devonshire, and Somerset. Very rare, and on a fair way to total extinction, so far as the British Isles are concerned.

Materials.—Twigs, heather-stalks, straws, and dry grass, sometimes wool, scantily and loosely lining the slight hollow chosen for the reception of the eggs.

Eggs.—Four to six. Very pale bluish-white, said to be sometimes marked with a few spots of pale reddish-brown. Average size about 1·65 by 1·4 in.

Time.—May.

Remarks.—Migratory, arriving in April and leaving in October. Notes : something like those of the Kestrel, but feebler and more querulous, according to Mr. Saunders. Local and other name : Ash-coloured Harrier. Sits lightly.

HAWK, SPARROW. *See* SPARROW HAWK.

HAWFINCH.

Description of Parent Birds. Length about seven inches. Bill of medium length, nearly conical, very thick at the base, and of a dusky blue colour. Irides grey. Crown and sides of head dull yellowish or orange brown; back and sides of neck ash colour. Back, smaller wing-coverts, and scapulars chestnut-brown. Some of the middle wing-coverts are white; wing-quills black, glossed with blue; some of them are of curious appearance, suggesting that they have been clipped at the tips so as to form battleaxes or billhooks. Rump and upper tail-coverts light orange-brown; tail-quills black, the outer ones being tipped and to some extent edged with white; middle greyish-brown, tipped with white. The feathers round the base of the beak, eyes, and on the throat are black; breast and belly pale rust colour; vent and under tail-coverts dull white. Legs, toes, and claws pale brown.

The female is less brilliant, and her colours are more mixed.

Situation and Locality.- In old lichen-covered hawthorn bushes; on the horizontal branches of oaks, heads of pollards; in holly bushes, firs, fruit, and other trees, at varying heights, in gardens, orchards, timbered commons, and plantations, pretty generally, though not commonly, in all the counties south of the six northern. I have met with it on Surrey commons, but never saw its nest in any part of Yorkshire, where it is said to breed.

Materials. Twigs, fibrous roots, and grass, mixed with lichens, and lined internally with fine fibrous roots, grass, and hair; somewhat loosely constructed.

Eggs.—Four to six, pale olive-green, varying to pale reddish-brown, or greenish-grey, spotted with blackish-brown, and irregularly streaked with dusky grey. Size about ·95 by ·75 in.

Time.—May.

Remarks.—Resident, although its numbers are increased in winter by Continental arrivals. Notes: call, rendered by Bechstein as an unpleasant *itszip*, uttered incessantly; song, a light jingle, with some clearer, shrill, and harsh notes like *irr*. Local and other names: Grosbeak Haw, Grosbeak, Common Grosbeak, Black-throated Grosbeak. Sits closely.

HEDGE SPARROW. *Also* HEDGE ACCENTOR.

Description of Parent Birds. — Length about five and a half inches; bill of medium length, almost straight, light brown at the base, and darker at the tip. Irides reddish-brown. Crown and nape dull bluish-grey, streaked with brown; back and

J

wings dusky brown, the feathers being edged with
reddish-brown; wing-quills dusky brown; tail-coverts
olive-brown; tail-quills dusky brown and slightly
forked; chin, throat, sides of neck, and upper
parts of breast dark bluish-grey; breast and belly
buflish-white; sides pale yellowish-brown, streaked
with a darker tinge of the same colour; vent
and under tail-coverts pale tawny brown; legs
and toes dark orange-brown; claws black.

The female is duller in plumage, with more
markings on the head and sides, and is a trifle
smaller.

Situation and Locality.—Hawthorn hedges are
favourite situations; the nest may, however, be
found in all kinds of low bushes, such as furze,
gooseberry, briars, brambles, and nettles. I met
with three nests, containing eggs, close together
amongst the black-currant bushes of a small garden
last year in Westmoreland; and remember on
one occasion finding one quite on the ground in
Yorkshire; and one, which was somewhat bulky in
appearance, in a bundle of pea-sticks in the corner
of a Surrey garden. Its size aroused my curiosity,
and I found that a new nest had been built on
the top of an old one, which contained three addled
eggs. Found pretty generally throughout the United
Kingdom, with the exception, perhaps, of the islands
lying to the north, and some of the smallest and
bleakest of those to the west of Scotland. Our
illustration is from a photograph taken in Hertfordshire, and shows a typical situation.

Materials.—Slender twigs (sparingly used and
sometimes entirely absent), roots, moss, and dry
grass, with an inner lining of wool, hair, and
feathers. I have on several occasions seen nests
made entirely of moss and cowhair.

HEDGE SPARROW.

Eggs.—Four to six, of a beautiful unmarked
turquoise-blue. Size about ·77 by ·6 in.
Time.—March, April, May, and June, sometimes
as late even as July.
Remarks.—Resident. Notes, a low plaintive
cheep-cheep, and a cheerful, though not long
sustained, song. Local and other names: Hedge
Accentor, Shufflewing, Hedge Warbler, Dunnock,
Hempie. Sits closely, and slips away without
demonstration.

HERON, COMMON.

Description of Parent Birds. — Length about
thirty-six inches. Beak long, straight, strong,
pointed, and dusky in colour, except at the base
of the under mandible, where it is yellowish. Irides
yellow. Forehead, crown, and cheeks, greyish-
white. On the back part of the head the feathers
are elongated into a kind of plume, and are bluish-
black or dark slaty-blue in colour. Upper surface
of body and wing-coverts bluish-grey; wing-primaries
black; tail-quills cinereous. Neck white, adorned
with large longitudinal elongated spots of dark
bluish-grey in front. On the lower part of the
neck the feathers are elongated, and hang loosely
down. Breast, belly, thighs, and under-parts greyish-white, streaked with black. Legs and toes
dirty yellowish-green; claws black.

In the female the plumes are shorter, and her
colour duller and less distinctive.

Situation and Locality.—On the tops of high
trees, ledges of cliffs, and in some places even on
the ground. The bird has been known to breed in
at least forty-one counties of England and Wales,
and does so in various suitable parts of Scotland

HERONS' NESTS.

(The lower nest in the circle is that of a ROOK with the hen sitting on her eggs.)

and Ireland. Our illustrations are from photographs taken in Wanstead Park, by the kind per-
mission of the Conservators of Epping Forest.

Materials.—A liberal collection of sticks and
twigs, lined with turf, moss, fibrous roots, and sometimes wool or rags, according to some authorities.

Eggs.—Three to five, pale blue with a tinge of
green. Size about 2·5 by 1·7 in.

Time.—January, February, March, and April.
The two first months only in exceptionally fine,
open seasons.

Remarks.—Resident. Notes, harsh, short, and
guttural. Some naturalists describe the alarm note
as *frank*, *frank*, *cronk*, but it sounds to me like
garowk, *garowk*, *garowk*. Local and other names:
Hearinsew, Hern, Heronshaw. Gregarious, as many
as eighty nests having been known in a single tree.
The birds return to the same place (called a
Heronry) year after year. Not a very close sitter,
as a rule. However, I have known individual
birds sit quite still whilst the trunk of a tree in
which their nests were situated has been violently
struck with a stick.

HOBBY.

Description of Parent Birds. — Length about
twelve inches. Beak short, much curved, and
bluish horn colour. Bare skin round the base of
the beak of a greenish-yellow colour. Irides dark
brown. Crown, nape, back, and wings greyish-
black, the feathers being edged with buffy-white.
Wing-quills almost black, bordered with light grey.
Tail-quills greyish-black, barred and tipped with a
lighter tint, except the two middle feathers, which

are uniform greyish-black. Chin and sides of neck white; cheeks black; breast and belly yellowish-white, streaked broadly with brownish-black; thighs, vent, and lower tail-coverts rusty red. Legs and toes yellow; claws black.

The female resembles the male, but is larger, and the spots on her breast are more conspicuous. In young birds of both sexes the plumage on the upper surface of the body is tinged with red, but this gradually gives place to bluish-grey with age.

Situation and Locality.—In high trees in woods and forests. It is almost as local as the Nightingale, and has not been reported as nesting further north than Yorkshire, except on one occasion in Scotland. It is a somewhat rare nesting species, but returns to a favourite haunt year after year.

Materials.—Some good authorities say that it does not build a nest of any kind, but simply adopts the old one of a Carrion Crow, Magpie, Woodpigeon, or that of some other Hawk; whilst others say that it builds a nest of sticks, moss, and hair.

Eggs.—Two to three, very rarely four, short and oval in form. Yellowish, dingy or bluish-white in ground colour, much suffused, mottled, and spotted with reddish and yellowish-brown. The eggs are similar in coloration and variety to those of the Merlin and Kestrel, and as the nesting situations are similar to those of the latter bird nothing short of a sight of the parents can settle the point with certainty. Size about 1·7 by 1·35 in.

Time.—May and June.

Remarks. — Migratory, arriving in April and leaving in October. Notes: alarm, a shrill chattering. Local and other names: none. Sits lightly, according to Mr. Dixon, and fairly closely, according to Mr. Seebohm, and is demonstrative when intruded upon.

HOOPOE.

This bird has been known in past times to breed in several parts of England, but the persecution it underwent in King Solomon's time, when tradition says that its handsome crest was made of gold, has so increased, that a detailed description of the bird, its habits, nest, eggs, etc., seems unnecessary in a work of this character. I fear a breeding pair of birds are never again likely to escape the lynx-eyed gunner in this country.

JACKDAW. *Also* DAW.

Description of Parent Birds. — Length about fourteen inches. Beak of medium length, strong, nearly straight, and black. Irides greyish-white. Crown black with a purple sheen; nape and back of neck leaden-grey. Back, wings, upper tail-coverts, and tail black, glossed with blue, violet, and green. All the under-parts are dusky black. Legs, toes, and claws black.

The female is a trifle smaller than the male, and the grey on the back of her neck is less pronounced.

Situation and Locality.—Holes in cliffs, church steeples, towers, old ruins, barns, and hollow trees, pretty generally throughout the British Isles. Our illustration is of a nest in the ventilation hole of a stone barn. It was slightly drawn forward, and light reflected upon it with a looking-glass, in order to take the photograph. The largest colony I have ever met with is near Armathwaite Castle, in Cumberland.

Materials.—Sticks, straw, moss, feathers, wool, down, and all sorts of odds and ends the bird can pick up near at hand. In some situations no sticks or twigs are used, and I have examined nests made entirely of rushes from beginning to end.

Eggs.—Three to six, usually five. Pale greenish-blue or bluish-white, spotted, speckled, and blotched with dark olive-brown and ash-grey. The markings vary in their distribution, being sometimes evenly distributed and at others collected round the larger end. The ground colour and markings are also subject to considerable variation. Size about 1·45 by 1·0 in.

Time.—May and June.

Remarks.—Resident. Notes, *kae*, or *caw*, and *jack*. Local and other names: Daw, Kae, Jack. Gregarious, and a close sitter.

JAY.

Description of Parent Birds. —Length about fourteen inches. Beak rather short, nearly straight, strong, and dusky. Irides white, slightly tinged with blue. Crown greyish-white, spotted and streaked with black and purplish-buff; the feathers form a crest, which the bird can elevate or depress at pleasure. Nape and sides of neck, back, and scapulars purplish-buff. Wing-coverts composed of alternate bars of pale blue, sky-blue, and black. Greater wing-quills black, with greyish-white edges; secondaries deep black, marked with a white patch on the upper half; rump white; tail dusky. From the gape, backward and downward, runs a moustache-like black dash; throat dirty white; breast pale purplish-buff; belly, vent, and under tail-coverts nearly white. Legs, toes, and claws brown.

The female is very similar to the male in appearance.

Situation and Locality.—In a tall thick bush, hedgerow, or young tree; sometimes in evergreens, such as the yew and holly; in woods and plantations with a thick undergrowth. Throughout the British Isles, nowhere very plentifully, but scarcer in some districts than others.

Materials. — Sticks, small twigs, mud, fibrous roots, and grass. Well built, as a rule, and somewhat like a large blackbird's nest.

Eggs.—Five to seven. Ground colours dusky green, tinged with blue, thickly spotted and freckled with light olive-brown. The markings are generally uniformly distributed, but are sometimes confluent at the larger end, where there are occasionally several irregular blackish-brown lines. Size about 1·25 by ·9 in.

Time.—April and May.

Remarks.—Resident. Note, a harsh, rapidly-delivered kind of chatter, sounding like *rake, rake.* Local and other names: Jay Piet, Jaypie. Not a very close sitter.

KESTREL.

Description of Parent Birds. — Length about thirteen inches. Bill short, much curved, and lead-coloured. Bare skin round the base of the beak, yellow. Irides dark brown. Head and nape of the neck ash-grey, under the eye is a dusky streak. Back, scapulars, and wing-coverts brownish-fawn colour, spotted with black; wing-quills black, edged

with grey; tail-feathers ash-grey, with a broad
black bar near the end, which is tipped with
white; under-parts light rust colour, spotted and
streaked with black; thighs, vent, and under tail-
coverts unspotted. Legs and toes yellow; claws
black.

The female is about two inches longer. Her
head and tail are reddish-brown, also the back,
which is duller than that of the male. On the
head are some dark streaks, and the back is barred
with bluish-black. The tail is very evenly and
prettily barred with black from the base to very
near the end, where the bars become broader.
Under-parts fainter than in the case of the male.

Situation and Locality.—On ledges and in
crevices of sea cliffs and inland crags and precipices;
holes in trees, towers, old ruins, church steeples,
and even dove-cotes have been utilised; also in
deserted nests of Crows, Magpies, and Sparrow
Hawks throughout the United Kingdom. The
nest represented was situated on the stump of a
tree growing horizontally, as near as possible, from
the crevice of a Highland precipice.

Materials. — Generally none at all, a cavity
being scratched in the soft earth and leaves in a
crevice or nook, which soon becomes plentifully
besprinkled with castings. Sticks, grass, and wool
are said to be sometimes used, but I never met
with either.

Eggs.—Four to seven, generally five to six; dirty
creamy-white in ground colour, thickly blotched
and clouded with reddish-brown. Very variable.
In some the ground colour is light brown, darkening
towards the larger end, blotched and spotted with
a darker shade. The colour in all varieties is, as
a rule, most abundant at the larger end. Size

YOUNG KESTRELS.

about 1·55 by 1·25 in. Indistinguishable from the eggs of the Hobby, and a sight of parent bird only can settle identity.

Time.—April and May.

Remarks.—Migratory, although a few specimens remain through the winter in Southern England. Note, a chattering kind of scream. Local and other names: Windhover, Staengall, Stannel Hawk, Stannel Hoverhawk, Stonegall, Creshawk, Standgale. A pretty close sitter.

KINGFISHER.

Description of Parent Birds.—Length about seven inches. Bill long, strong, straight, and black, except at the base of the under mandible, where it is orange. Irides hazel. Crown, nape, back, wings, rump, upper tail-coverts, and tail dark, greenish-blue; the head and neck are barred with brilliant azure blue. The wing-coverts are spotted with the same colour, which is prominent on the middle of the back, rump, and upper tail-coverts. Wing-quills dull greenish-black, greenish-blue on the outer webs, and reddish-brown on the outside edges of the inner, except at the tips, which are dull black. From the base of the upper mandible to the eye, and thence to the ear-coverts, chestnut. Chin and throat dirty white, slightly tinged with rust colour. Breast, belly, sides, vent, and under tail-coverts, beautiful chestnut; duller on the last two mentioned parts. Legs and toes pink ; claws, brownish-black.

The female has a shorter beak, and is slightly duller in her plumage.

Situation and Locality.—A hole in river or other bank, generally well hidden by some overhanging

KINGFISHER.

piece of earth, undermined by the action of the
water; occasionally in the side of a sand-pit, or,
more rarely still, in a hole in a wall. The hole
is from one to three or four feet in length, sloping
upwards, and ending in a rounded chamber. The
bird, in some instances, excavates it, and in others
adopts the old nest of a Sandmartin, or even a rat's
hole. General, but not numerous, in most suitable
parts of the United Kingdom. The bird is said to
be most abundant in the neighbourhood of Oxford,
and absent from the most northern parts of Scotland. The illustration is from a photograph taken
on the Mole in Surrey.

Materials.—Fish bones in variable quantities, and
the dry mould of the hole.

Eggs.—Five to eight, sometimes as many as ten.
Of a beautiful pink colour before being blown, on
account of the yolk showing through, but snowy-
white and glossy afterwards. Size about ·9 by ·75 in.

Time.—February, March, April, May, June, and
July.

Remarks.— Resident. Notes, a shrill squeal,
which never fails to attract the naturalist's attention. Local and other names: Halcyon. Sits
closely, and generally betrays the whereabouts of
its nest by the white droppings near the entrance
to the hole.

KITE.

Description of Parent Birds.—Length about
twenty-six inches. Beak shortish, hooked at the
tip, strong, and horn-coloured. Bare skin round the
base of the beak, and irides yellow. Head and
neck light grey, streaked with cinerous brown; back
and wing-coverts dusky, bordered with rusty red.

Wing-quills dusky black, some of the inner ones being edged with white on the interior webs. Upper tail-coverts rusty red; quills rusty brown, barred on the inner webs with dusky brown. The tail is much forked. Under-parts rusty brown, whitish on the chin, throat, and under tail-coverts, and streaked with dusky brown, except on the last-named part. Legs and toes yellow; claws black.

The female is somewhat larger, and is said to be greyer about the head and redder beneath the body. However, some ornithologists say that she is less red than the male.

Situation and Locality.—In the forked branch of a tree, or on several branches close to the trunk, at varying heights, in the densest parts of woods and forests. Some authorities seem to be of opinion that the bird has ceased to breed in the British Isles, but Mr. Dixon says that it still does so in some favoured localities of Wales and Scotland.

Materials.—Sticks and twigs are used liberally for the outsides, and the foundation lined with moss, wool, grass, and any rubbish the bird can pick up, such as bits of paper and rags.

Eggs.—Two to four, generally three. Greyish dirty white, spotted, blotched, and streaked with dull red and brownish-yellow, with underlying markings of greyish-lilac. The markings are generally most numerous at the larger end. Subject to considerable variation. Size about 2·25 by 1·75 in.

Time.—May.

Remarks. — Resident. Note, a shrill shriek, known in some localities as a " *whew.*" Local and other names: Glead, Fork-tailed Kite, Fork-tailed Glead, Gled or Greedy Gled, Puttock, Crotchet-tailed Puttock, Glade. Sits pretty closely, and it is said will defend its nest when in danger of having it robbed.

K

KITTIWAKE. *Also* KITTIWAKE GULL.

Description of Parent Birds.—Length about fifteen and a half inches. Bill of medium length, slightly curved downward, and greenish-yellow in colour. Irides dusky brown. Head and neck white; back and wings pale grey, the longest quills of the latter tipped with black. Tail-coverts and quills white. Chin, throat, breast, belly, vent, and under tail-coverts snowy white. Legs, toes, and membranes dusky. The female is similar to the male.

Situation and Locality.—In crevices and on ledges of rock, at various heights above the sea. It will be noticed in the photograph that some of the ledges are so small in area that the birds have to sit in peculiarly uncomfortable positions in order to cover their eggs. Colonies inhabit a great number of suitable breeding-places round our coasts.

Materials.—Heath, dry seaweed, and dead grass, somewhat carelessly arranged.

Eggs.—Two to four. Some authorities say that two is the most general number; others three. Ground colour varies from stone-yellow to buffish-brown, sometimes shaded with blue, blotched and spotted thickly with ash-grey, light brown, reddish or umber brown. Very variable, both in ground colour and markings. Size about 2·15 by 1·6 in. The size of the eggs, their large markings, and the situation of the nest prevent confusion with those of any other gull.

Time.—May and June.

Remarks.—Resident, but subject to much local movement. Notes, *kitt-aa*, *kitti-aa*, which sound like " get away, get away." Local and other names : Tarrock, Annet, Hacklet or Hacket Gull, Waeg, Mackerel Bird. Gregarious, and a close sitter.

KITTIWAKES ON THEIR NESTS AT THE FARNE ISLANDS.

LANDRAIL. See CRAKE, CORN.

LAPWING. Also PEEWIT.

Description of Parent Birds. — Length about twelve inches. Beak somewhat short, straight, and black. Irides hazel. Forehead, crown, and long, narrow, upturned crest black, glossed with green. From the base of the beak a dirty white line passes over the eye and round under the crest. A streak of black runs under the eye and round to the nape of the neck, which is brown, mixed with white. Back and wing-coverts and scapulars brownish-green, glossed with purple and blue. Primaries black, with white spots at the ends of the first three or four; secondaries white on the basal half. Upper tail-coverts reddish-chestnut. Basal half of tail white; lower half black. Chin, throat, and upper half of breast black; lower half of breast, belly, and vent white; under tail-coverts pale rust-colour. Legs and toes dull fleshy pink; claws black.

The female has the crest much shorter, and is less bright in her colour.

Situation and Locality.—On the ground in rough pasture-land, marshes, fallow fields, and other suitable places throughout the British Isles. I know one place within seven miles of the bricks of London where the bird breeds year by year. Our illustration was procured on the Westmoreland hills.

Materials.—A few bits of dry grass, rushes, or moss, used as a lining to the depression in which the eggs are laid.

Eggs.—Four, although five have been reported.

LAPWING.

The latter number must be very rare, for I have found a great many nests, but never once saw more than four; and know several gamekeepers who collect eggs for table use every spring, and they do not recollect ever meeting with a nest containing more. Dirty olive-green, blotched and spotted all over with blackish-brown. Sometimes the ground colour is light buff or buffish-brown, of various shades. The markings are generally most numerous round the larger end. Size about 1·85 by 1·35 in.

Time.—April and May, and sometimes as late as June.

Remarks.—Resident, though subject to southern movement in winter. Notes, *peewit*, the first syllable long-drawn when used as a call-note, and short when the bird is alarmed. Local and other names: Peewit, Green Plover, Peeweep, Tufit, Crested Lapwing. Sits lightly. A great deal has been written to prove that the bird rises straight off its eggs by some observers, and that it runs for some distance before taking wing by others. My own experience has been that the bird is exceedingly quick of eye and ear, and if the intruder be discovered at some distance, the bird runs before rising; but if suddenly alarmed at close quarters, it will rise straight off its nest. A good way to find the Lapwing's eggs is to creep very quietly up behind the wall or hedge of a field or pasture in which the birds are known to breed, and then show oneself suddenly, and mark where the birds rise from.

LARK, SKY. *See* SKYLARK.

LARK, WOOD. *See* WOODLARK.

LINNET.

Description of Parent Birds.—Length about five and a half inches. Bill short, broad at the base, sharp-pointed, and bluish-grey in colour. Irides hazel. The plumage is subject to great variation. Forepart and top of head brownish-red; rest of head, back, and sides of neck brownish-grey; back and upper wing-coverts deep rufous-brown; wing-quills dusky, edged with white; upper tail-coverts dark brown; tail-quills brownish-black, edged with white, except the two centre feathers. Chin and throat and under-parts dirty reddish-brown, the red being brightest on the breast and very variable in its intensity, some having very little of it present. Legs, toes, and claws brown.

The female is a trifle smaller, lacks the red on the top of the head and breast, the feathers on her head, neck, back, and rump being dark-brown, edged with a paler tint of the same colour; her under-parts are dull yellowish-brown, streaked with dark brown. She is said to have less white on the wing-quills.

Situation and Locality.—Furze bushes, white and black thorn bushes, heath, juniper bushes, amongst tall heather, and even on the ground upon rare occasions. Our illustration is from a photograph obtained on a common near Lowestoft. It is one of two or three which I found, and sent my brother specially down to photograph. He has, however, exposed the nest, and raised the eggs a trifle too much. The nest is met with on furze-clad sides of hills, commons, and rough uncultivated lands covered with heather, furze, and ling, throughout the British Isles.

Materials.—A few small twigs, fibrous roots, dry grass-stems, moss, and wool, with an inner lining of hair and feathers, sometimes with rabbit or vegetable down.

Eggs. — Four to six, greyish-white, slightly tinged with blue or green, speckled and spotted with purple, red, and reddish-brown; the spots are generally most numerous round the larger end of the egg. They closely resemble those of several other birds, such as the Greenfinch, Goldfinch, and Twite, and can only be distinguished with certainty by watching the parent birds on to or off their nests. Average measurement about ·72 by ·52 in.

Time.—April, May, and June, sometimes as late as July, and even August.

Remarks.—Migratory in bulk, but resident in small numbers. Notes: song, soft and low, mixed with some sweet and shrill notes. Local and other names: Brown Linnet, Common Linnet, Grey Linnet (a name referring to the young bird before the first moult), Whin Linnet, Red-breasted Linnet, Rose Linnet, Greater Redpole, Red Linnet, Linnet Finch, Red-headed Finch, Lintie, Linwhite. Sits close.

LINNET, MOUNTAIN. *See* TWITE.

MALLARD. *See* DUCK, WILD.

MAGPIE.

Description of Parent Birds. —Length about eighteen inches, more than half of which is accounted for by the bird's abnormally long tail.

LINNET

Bill of medium length, slightly curved downward, and black. Irides hazel. Head, neck, back, wings, tail-coverts and quills black, glossed with green. purple, and blue, according to the light upon them. The scapulars, and part of the inner webs of some of the primaries, are white. The tail is very much wedge-shaped. Chin, throat, and upper breast black; lower breast, belly, and sides white; thighs and under tail-coverts black. Legs, toes, and claws black.

The female is not so large, nor is her plumage so brilliant as in the case of the male.

Situation and Locality.—In trees and thorn bushes at varying heights from the ground. Instances are on record of the bird having made its nest even in a gooseberry bush. It is sometimes partial to situations near the habitations of man, and a small clump of trees or thorn bushes will suit its purpose quite as well as a big wood or plantation, apparently, for I have as often found it in one as the other. Pretty generally over the British Isles. except Orkneys, Shetlands, and outer Hebrides. Our illustration was obtained in Westmoreland.

Materials. — Dead thorn-sticks, brambles, and twigs interlaced; those forming the foundation of the nest are knitted together with liberal quantities of clay and mud. The nest is bulky. domed, and spherical, with a hole on the side and nearish the top. It is lined internally with fibrous roots, and is such a substantial affair that I once saw a gamekeeper shoot at one from the foot of a fir tree some fifty feet in height without breaking one of the six eggs inside.

Eggs.—Six to eight or nine, the first number being the most general. Dirty bluish-green, or

MAGPIE.

yellowish-brown, spotted, freckled, and blotched all
over with grey and greenish-brown.   Size about
1·35 by ·95 in.

Time.—March, April, and May.

Remarks.— Resident.   Notes, chattering.   Local
and other names: Pyet, Madge, Mag, Maggie,
Pianet, Hagister.   I have known the bird sometimes sit closely when not far advanced in incubation, and at others lightly ; something, I am inclined
to think, depending upon the height of the nest
from the ground.

MARTIN.   *Also* MARTIN, HOUSE.

Description of Parent Birds.—Length about five
and a quarter inches.   Bill short, flat, and wide
at the base, and black.   Irides brown.   Crown,
nape, and sides of the head, back and wing-coverts,
dark steely-blue.   Wing-quills dull black ; rump
and upper tail-coverts white ; tail-quills dull black.
The tail is forked, but so much less so than that
of the Swallow that, apart from any difference of
coloration, it easily distinguishes the bird.   Chin,
throat, breast, belly, and under-parts generally,
white.   Legs and toes short, and almost hidden by
a profusion of fine, soft, white feathers.   Claws grey.

The female is very similar indeed in appearance,
although her plumage is perhaps not so bright.

Situation and Locality.—Under the eaves of
houses, stables, barns, and other buildings, angles
of windows, under the projecting " through " stones
of barns ; in nooks and corners of rocks and sea
cliffs.   Our illustration is from a photograph, taken
whilst two of the young ones had their heads out
awaiting anxiously the return of their parents.
I know a small stable in Surrey, under the eaves

HOUSE MARTIN.

of which, and principally on the side with a
south-east aspect, I have counted forty-seven nests
several years in succession. My brother and I went
down in 1894 specially to photograph the building,
and to our great disappointment there was not a
single nest under its eaves! The owner informed
us that it was the first time he had noticed the
absence of the Martins for twenty-five years, and
attributed it to the droughty summer of 1893
having made suitable building materials difficult to
procure, and the unbearable persecution and robbery
of the Sparrows, General over the British Isles.

Materials.—Clay or mud made into pellets and
cemented together until they form a kind of shell
like the half of a deep basin, fixed close up under
an eave or projecting object, with a small elliptical
hole at the top, and generally on one side. It is
lined internally with bits of straw, hay, and
feathers.

Eggs.—Four to five, rarely six ; white and unspotted, the yoke giving them a slight pinky tinge
before they are blown. I see by my list of rare
natural history notes, culled from that admirable
paper, that one or two observers have recorded
in the *Field* the finding of rust-red spotted specimens ; but I have never met with a single egg
showing any inclination in this direction out of many
scores of nests examined. Size about ·8 by ·52 in.

Time.—May, June, July, August, and even as
late as September.

Remarks. — Migratory, arriving in April and
leaving in September and October; although individuals are frequently reported in November and
even December. Notes : call, something like *spitz*,
but very difficult to represent. Local and other
names : Window Martin, Window Swallow, Eave

Swallow, Martlet. Gregarious, as a rule. Sits closely.

MARTIN, HOUSE. *See* MARTIN.

MARTIN, SAND.

Description of Parent Birds.—Length about four and three-quarter to five inches. Bill short, slightly turned down at the tip, and brownish-black. Irides hazel. Head, back of neck, and all upper-parts, including wings and tail, brownish-black or mouse-colour. Throat and breast white, the latter having a band of lightish-brown running across it. Belly and under-parts white. Legs and toes reddish-brown. The brown tinge of the upper-parts and the smaller size readily distinguish this bird from the House Martin.

The female differs very slightly from the male.

Situation and Locality.— At the extremity of a tunnel, dug by the bird's own exertions. It varies in length from eighteen inches to three or four feet, and generally slopes upwards from the entrance to the little round chamber in which the nest is situated. The gallery is about two inches in diameter, generally crooked. In the banks of rivers, sand-pits, railway cuttings, and lanes with high sandy banks. Our illustration is from a photograph taken near Nutfield, Surrey, and is of double interest—firstly, it is quite away from any water, and secondly, every nest was taken possession of in 1894 by Sparrows. These little birds breed all over the country, penetrating to the distant Orkney and Shetland Islands.

Materials. — Straw and grass stems, with an

inner lining of feathers. The whole is very loosely put together, and I have met with specimens with no feathers at all, and but very few hay straws whereon the eggs were laid.

Eggs.—Four to five, seldom six; pure white when blown. The shell is so thin and semi-transparent that the yolk shows through and gives the egg a pinky tinge. Size about ·7 by ·48 in.

Time.—May, June, and July.

Remarks.—Migratory, arriving in this country in March and April, and leaving in September and October. Call-notes loud and harsh, something like *share*, according to Naumann, but very difficult to represent in letters. Local and other names: Pit Martin, Land Swallow, Bank Swallow, River Swallow, Bank Martin. Gregarious. Sits closely.

MERGANSER, RED-BREASTED.

Description of Parent Birds.—Length about twenty-two inches. Bill rather long, sharp, straight, and red, except the upper part, which is brownish. Irides red. Head and a little of the upper part of the neck glossy green, the feathers on the back of the head being lengthened. A line of black runs from the back of the head down behind the neck to the upper part of the back, which is also black; lower part of back, rump, and upper tail-coverts grey. Tail-quills brownish-grey. Wings a mixture of dark brown, white and black on the upper-parts; primaries brownish-black; middle half of sides of neck white; breast rusty red, spotted with black on the front; on the sides, in front of wing-points or shoulders, are a few white feathers edged broadly with black. Breast, belly, and under

SAND MARTINS' NESTING HOLES.

L

tail-coverts white. Legs, toes, and webs deep orange, tinged with brown; claws black. At the latter end of May the head and neck turn from glossy green to dull brown, and the rusty red on the breast disappears.

The female is slightly smaller; her head, neck, and the whole of her upper-parts are of varying shades of brown, with two white bars on the wings. Front of neck white, mottled with light reddish-brown; under-parts white. Both sexes subject to variation of colour.

Situation and Locality.—On the ground, under bushes, banks, projecting ledges of rock; amongst heather and brambles; occasionally in holes in trees, in rabbit-holes, holes and crevices of rocks; on small islands in lakes, on the shores of lakes, generally not far from the water. In the North of Scotland, Orkneys, Shetlands, Hebrides, and in Ireland. Our illustration was procured in the Highlands, where we met with several nests, some of which had been been destroyed by Hooded Crows.

Materials.—Dead grass, roots, and rushes, in scanty quantities, lined with tufts of down from the bird's own body. These are light greyish-brown, with pale centres and tips. Sometimes no materials whatever except down are provided.

Eggs.—Six or seven to eleven or twelve, olive-grey to buffish-grey, somewhat similar to those of the Scaup. Size about 2·6 by 1·7 in.

Time.—End of May, June, and beginning of July.

Remarks.— A winter visitor; numbers, however, stay and breed with us. Local and other names: Red-breasted Goose, Sheld Duck, and Spear Wigeon (the latter two names only applied to the bird in Ireland). A very close sitter.

RED-BREASTED MERGANSER.

MERLIN.

Description of Parent Birds.—Length about ten inches. Beak short, much curved, and bluish-horn colour. Bare skin round the base of the beak yellow. Irides dark brown. Crown bluish-grey, marked with black streaks along the shafts of the feathers. Cheeks and upper part of neck rusty brown, marked with blackish streaks ; back, scapulars, wing-coverts, and rump bluish-grey, each feather having the shaft black ; wing primaries black. Tail-quills, like the back, barred with a darker hue, and tipped with white. Chin and throat nearly white ; breast, belly, sides, and thighs rusty red, streaked with dusky brown ; vent and under tail-coverts pale rust-colour. Legs and toes yellow ; claws black.

The female is about two inches longer than the male. The whole of her upper-parts are dark liver-brown, the feathers tipped with rusty red and having dusky shafts. Tail-quills, like the back, barred with light yellowish-brown. Under-parts pale brownish-white, with broad, dusky brown streaks.

Situation and Locality.—On the ground, amongst deep heather and ling, or scattered rocks ; on open moors, heaths, and rough sheep-pastures. It is said to be occasionally found in trees and on cliffs, but I have never seen one in either situation. In the wild moorland parts of the North of England, Wales, Scotland, and Ireland. Our illustration is from a photograph taken on the hills between Westmoreland and Yorkshire. The nest was in deep heather on a sloping hillside, commanding every aspect of approach. It was evidently a favourite site, for the gentleman who showed it to us in 1894 said that a brood had been reared at the same

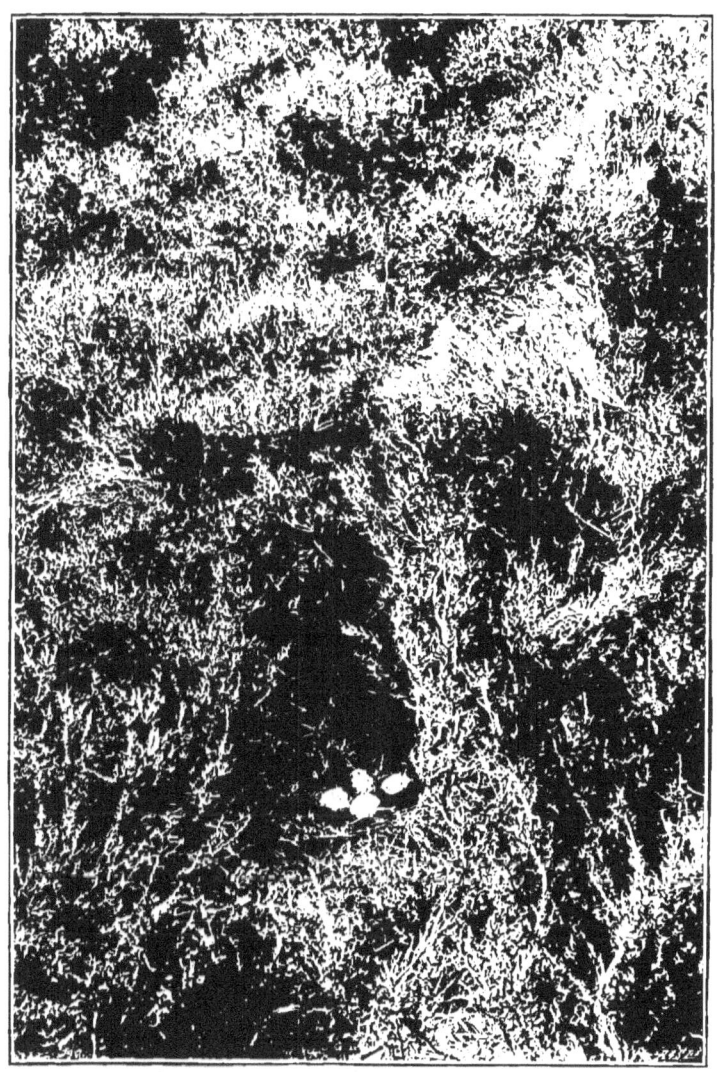

MERLIN.

place the year before; and I flushed the female close to the place this year, but was unable to find her nest. He showed us three knolls, each about fifty yards from the nesting site, upon which the old birds plucked the prey before taking it to their young. They brought Moor Poults (young Grouse), Green Plovers, Meadow Pipits, mice, and young Snipe.

Materials.—A few twigs or sprigs of heather, grass, or moss, generally next to nothing. The one photographed was in a very slight depression, and contained only a few dead heather sprouts.

Eggs.—Three to six, generally four or five, creamy-white, so thickly covered with spots, blotches, or clouds of dark reddish-brown, as to almost completely hide the ground colour. Sometimes the markings consist of small dots closely and thickly scattered over the whole surface, and in these the ground colour becomes more apparent. Size about 1·6 by 1·2 in. Only distinguishable from those of the Hobby and Kestrel by the situation of the nest.

Time.—May and June.

Remarks.—Resident, but subject to a southern migration in October. Notes, a shrill, chattering cry. Local and other names: Blue Hawk, Stone Falcon. Sits lightly.

MOORHEN. *Also* WATERHEN.

Description of Parent Birds.—Length about thirteen inches. Bill of medium length, nearly straight, greenish-yellow at the tip, and red at the base and for some distance up the naked forehead, where the coot is white. Irides reddish-hazel.

MOORHEN.

MOORHEN.

Head and neck dark bluish-grey. Back, wings, rump, and upper tail-quills dark olive-brown. Breast and sides dark bluish-grey ; belly and vent grey ; flanks streaked with white. Under tail-coverts black. The under tail-quills are white. Legs and feet greenish-yellow ; claws rather long and brown. Above the knee is a broad garter of red.

The female is rather larger, and brighter in the coloration of her plumage than the male.

Situation and Locality.—Generally on the ground, amongst flags, rushes, reeds, low bushes growing from the water ; reeds and coarse aquatic plants growing in and on the banks of rivers, small streams, canals, ponds, lakes, and reservoirs. I have, however, met with it at considerable heights above the water, amongst rubbish left by an abnormally high flood in a tree. Our illustrations were procured on a small island in the middle of the River Mole. Common throughout England, Wales, Scotland, and Ireland.

Materials. — Flags, reeds, rushes, and grass. Generally in small quantities where the situation is dry, but often in a fair-sized matted mass where its base is in the water.

Eggs.—Seven to ten. Buffish-white or rusty buff, spotted and speckled with reddish-brown of various shades. The markings are not very large, or profusely distributed. Size about 1·7 by 1·2 in.

Time.—March, April, May, June, July, and even as late as August sometimes.

Remarks. — Resident, and partially migratory. Notes : call, *crek-rek-rek.* Local and other names : Water Hen, Marsh Hen, Moat Hen, Gallinule. Not a very close sitter, slipping quietly off the nest and instantly hiding on the approach of any intruder.

NIGHTINGALE.

Description of Parent Birds.—Length about six inches. Bill of medium length, nearly straight, and brown. Irides hazel. Head and upper parts of body uniform tawny brown. Wing and tail-quills brown, edged with rust-colour. Chin, throat, and all under-parts greyish-white, tinged with brown on the breast, and reddish on the under tail-coverts. Legs, toes, and claws brown.

The female is rather smaller in size, but otherwise closely resembles the male.

Situation and Locality.—In natural declivities on the ground, on little banks at the foot of trees, amongst exposed roots at the bottom of hedgerows, under the shelter of ferns or weeds. I have known one at some little height from the ground, amongst the dead weeds, twigs, and leaves that had been built up round the trunk of a tree for the purpose of hiding a gunner in pursuit of Wood Pigeons. In woods, groves, small shady copses, plantations, quiet gardens, and commons with clumps of hazel briars and brambles growing thereon. It is peculiarly limited in its habitat, as a rule going no farther north than Ripon in Yorkshire, and no farther west than the Valley of the Exe, although individuals have been met with beyond these limits, and there is reason to believe it is extending its range. Our illustration was obtained in Norfolk.

Materials.—Dry grass-stalks, leaves, moss, bits of bark and fibrous roots, lined inside with fine grass and horsehair.

Eggs.—Four to six, generally five. Uniform olive-brown or olive-green. Specimens have been met with occasionally of a greenish-blue colour,

...... the reddish-brown which colour upon the
greenish-blue produces dissolution. Sometimes the
......is disposed in a kind of cap at one or other
of the ends of the egg, or in streaks. Size about
·7 by ·5 in.

Time. — May and June.

Remarks. — Migratory, arriving in April and leaving
in August. Notes: call, *peer, peer,* and something
like *churr.* Seebohm rendering it *torr torr,* and
adding that it has a note of enjoyment represented
by a deep *coo.* Its song is uttered both by day
and night, and is exceedingly sweet and melodious.
Local and other names: Fernowl, ..., ... closely
and slips away with ......

NIGHTJAR. *(No illustration.)*

Description of Eggs and Young. — Length about ten
and a half inches. Bill very short, upper mandible
slightly curved downward, flexible, and almost black.
The gape is very wide, and furnished at the upper
side with a number of stiff bristles. The plumage
of the upper part of the bird consists of a beautifully variegated mixture of brown, buff, rust-red,
and white, streaked and sprinkled with grey. The
under-parts greyish and rust-brown, barred and
freckled with dark brown. There are a few white
markings round the throat. On the three first
quill-feathers of the wings is an oval white spot,
also on the two outside feathers of the tail. Legs
short, rough, and the tarsi feathered to below the knee.
Middle toe considerably longer than the rest, and
its claw upon it is serrated on its side; all colour
brown.

Situation and Locality. — On the ground beneath

furze bushes, brackens, heather, or quite in the
open on commons, heaths, open, bramble-covered
woods and copses in nearly all suitable districts
throughout the British Isles, but not very numerous
anywhere.   I have met with it most frequently
in the southern and eastern counties of England.
Our illustrations are from photographs taken on a
Suffolk common near Lowestoft.

Materials.—None, the eggs being laid on the
bare ground.

Eggs.—Two; ground-colour white, greyish-white,
or creamy-white, clouded, blotched, marbled, or
veined with dark brown, and underlying tints of
bluish lead-colour.   They are subject to great
variation, and often closely resemble flint pebbles
one may pick up on the beach with chalk adher-
ing to them; in fact, I have on more than one
occasion been deceived by one of these pebbles
lying under a furze bush.   Size about 1·25 by ·87 in.

Time.—May and June.

Remarks. — Migratory, arriving in May and
departing in September or October.   Note, *jar-
r-r-r-r-r* and *dee*, *dee*, said to be uttered on taking
flight.   Local and other names : Goatsucker, Dor-
hawk, Fen Owl, Nighthawk, Wheelbird, Jar Owl,
Churn Owl, Goat Owl.   Sits closely, trusting to
the wonderful harmony of her plumage with surrounding objects.

NUTHATCH.

Description of Parent Birds.—Length about six
inches; bill moderately long, strong, nearly straight,
sharp at the tip, and bluish-black, except at the
base of the lower mandible, where it is whitish.
Irides hazel.   Crown and all upper parts of body,

NIGHTJAR'S EGGS AND NEWLY-HATCHED YOUNG

including wing-coverts and part of tail, bluish
slate-grey. Wing-quills dusky, margined on the
outer webs with blue. Tail-quills, excepting those
mentioned above, black, tipped with grey, and
marked on either side with white. A black streak
passes from the base of the bill to each eye, and
thence down the side of the neck. Sides of head
and chin white; throat, breast, and belly buff;
sides and thighs dark rust-colour or chestnut; vent
white, marked with rust-colour. Legs, toes, and
claws light brown, inclining to yellowish.

The female is lighter coloured on her under-parts.

Situation and Locality.—In a hole in the trunk
or strong branch of a tree, old stumps, and occasionally in a hayrick or wall; at varying depths of
from three or four to twelve or fifteen inches. In
the south and midlands of England, also Wales;
rarely met with in Scotland and never in Ireland.

Materials.—Leaves, flakes of bark, dry grass,
and sometimes chips and *débris* when the bird is
obliged to enlarge the situation selected. The
bird has the peculiar habit of plastering up the
approach to its nest with clay if there be more
room than is necessary for its admission.

Eggs.—Five to eight or nine; pure white, spotted
with reddish-brown; sometimes blotched, the markings varying in distribution. If care is not exercised
the eggs are likely to be mistaken for those of the
Great Titmouse, but the character of the nest will
readily settle the point. Size about ·8 by ·57 in.

Time.—April, May, June, and July.

Remarks.—Resident. Notes: call, *whit, whit,
whit;* sometimes represented as *twi-twit, twi-twit.*
Local and other names : Woodcracker, Nutjobber,
Jarbird, Nuthack, Mudstopper. Sits very closely,
and hisses like a snake when disturbed.

ORIOLE, GOLDEN.

This bird, although a somewhat rare and accidental visitor to our shores, has, according to some authorities, bred in several parts of England. There are, however, sceptics who doubt this, and adduce, as a reason, that there is not a collector who can boast the possession of a British-laid specimen. Be this as it may, it is doubtful whether the bird will ever succeed in breeding in this country, on account of the eagerness with which the collector seeks after the skin of the male, whose attractive colours excite his cupidity.

OSPREY.

Description of Parent Birds. — Length about twenty-two inches. Beak short, much curved, and black ; naked skin round base of beak blue. Irides yellow. Crown and nape whitish, streaked with dark brown, the feathers somewhat elongated into a kind of crest. Back and wings dark brown, sometimes glossed with purple, ends of the latter black. Tail waved with two shades of brown ; chin and throat white, shaded with light brown across the breast ; belly, sides, thighs, and under tail-coverts white. Legs and toes blue, claws long, strong, much curved, and black.

The female is slightly larger, and has her head and breast more marked with brown, according to Mr. Seebohm.

Situation and Locality.—Near the top of a high tree, the summit of an inland crag, or on the highest point of some ruin upon an island or commanding promontory amongst the lonely lochs of the Highlands of Scotland. It is only known to breed in one or two counties (Inverness-shire, Ross-shire, and Galloway), and would undoubtedly have long ago been banished, had it not been for the strict protection extended towards it where it now builds.

Materials.—Sticks, twigs, turf, moss, and grass. The structure is of a huge character, and the top almost flat. The same site is used again and again with the utmost regularity.

Eggs.—Two to four, generally three; very variable and beautiful. The ground colour ranges from white to dull yellowish-white, handsomely marked with rich reddish-brown and light brownish-grey. Some examples are suffused with bright orange-red or purple. The blotches and spots are sometimes very thickly distributed, at others they form a zone round the larger end or are irregularly scattered over the entire of the egg. They (the eggs) also vary considerably in size. Average about 2·3 by 1·85 in.

Time.—May and June.

Remarks.—Migratory, arriving in April or May, and departing in September and October. Notes, *kai, kai, kai.* Local and other names: Eagle Fisher, Mullet Hawk, Fish Hawk. Sits lightly, according to Mr. Dixon, but pretty closely according to Mr. Seebohm.

OUZEL, RING. *See* RING OUZEL.

OUZEL, WATER. *See* DIPPER.

OWL, BARN.

Description of Parent Birds. — Length about fourteen inches. Beak short, much curved at the point, and pale grey. Irides black. Discs round the eyes white, with the exception of a little patch close to the eyeball on the inner side of each, which is rufous. The feathers, especially on the lower outside of each disc, are tipped with light rusty brown of varying shades. Crown and nape light buff, sprinkled with grey and spotted with dark brown and dirty white. Back, wings, and rump buff, with a lacework of grey, on which are more or less perpendicular lines of spots of dull black and dirty white; upper surface of tail-quills greyish-buff, crossed by fine darkish-grey bars; upper breast white, slightly tinged with buff; lower breast, belly, vent, and under tail-coverts white. Legs covered with white downy feathers, toes with short hairs; claws brown.

The female is distinguished by a few dark brown spots on her sides and belly.

Situation and Locality.—In hollow trees, church towers, barns, pigeon-cotes, crevices of rocks overshadowed by ivy, and old ruins; pretty generally throughout the British Isles, but scarce in the Highlands, Orkneys, and Shetlands. Our illustration represents a hollow tree at Shenley, in Hertfordshire, in which a pair of Barn Owls have nested for many years.

Materials.—None, usually, except the pellets of undigested parts of birds and mice; however, in some situations a few sticks or twigs and other materials, such as straws, wool, and hair in small quantities, are said to be used.

M

Eggs.—Two to six; white, without polish or markings of any kind. The bird commences to sit as soon as she has laid one or two eggs, and keeps on laying one or two at intervals, so that the young in the same nest may often be found in various stages of development. Average size about 1·6 by 1·25 in. Their smaller size distinguishes them from those of the Tawny Owl, and their situation from those of both species of Horned Owls.

Time.—April, May, June, and July, although young have been found as late even as December.

Remarks.—Resident. Note, a loud screech. Local and other names: White Owl, Hissing Owl, Church Owl, Madge Howlet, Jinny Oolet or Oolert, Screech Owl, Yellow Owl. Sits close, and is partial to old situations.

OWL, LONG-EARED.

Description of Parent Birds.—Length about fourteen inches. Beak short, much curved, and dusky horn colour. Irides orange-yellow. Facial discs dusky brown near the centre on the inside, and white towards the ends of the feathers; the outer sides of each disc are pale brown, ending in a line of darker brown. The ears or tufts of feathers on the head are about an inch and a half long, greyish-white on the inner edges, brownish-black in the centre, and dullish yellow on the outside edges. Crown, between horns, a mixture of the same colours. Nape, neck all round, and the upper portion of the back, dull yellow, streaked, longitudinally, with brownish-black. Lower back and wings yellowish-brown, marked with greyish-white, dark brown, and black. Upper side of tail rusty red, barred and speckled with dark brown; breast and belly

HOLLOW TREE IN WHICH A PAIR OF BARN OWLS
HAVE NESTED FOR SEVERAL YEARS.

greyish-white, mixed with pale brown, and streaked and barred with dark brown. Under tail-coverts, and feathers on legs and toes, pale yellowish-brown; claws same colour as beak.

The female is similar in plumage, but is said to be somewhat larger.

Situation and Locality.—The old nest of a Crow, Heron, Magpie, Wood Pigeon, or the disused drey of a squirrel, in plantations of firs, and in woods and forests containing evergreens sparingly, in suitable localities throughout the United Kingdom.

Materials.—None.

Eggs.—Three to seven, generally four or five. White, oval, and smooth. Size about 1·65 by 1·3 in. Not likely to be confused with those of any other bird except Ring Dove; but their number and the appearance of the layer will readily settle the point.

Time.—March and April.

Remarks.—Resident and also migratory. Note, a deep hoot. Local and other names, none. Sits very closely.

OWL, SHORT-EARED.

Description of Parent Birds.—Length about fifteen inches. Beak short, much curved, and blackish. Irides yellow. The radiating circle of feathers round each eye black in the centre, and lighter on the outer edges, mixed with reddish-brown, black, and white, especially the last-named colour, round the bill. On the top of the head are two tufts of feathers about three-quarters of an inch long, which the bird can erect or depress at pleasure. These are blackish on the outer webs, and whitish on the inner. Crown of the head, neck, back, and scapulars dusky, the feathers being bordered with light rusty brown. Wing-coverts dusky, marked with a number of yellowish-white spots; primaries light rusty brown, barred with blackish-brown. Tail-quills pale rufous, barred with dark brown. Under-parts buffish-white, streaked on the breast and belly with blackish-brown. Legs and toes feathered and pale buffish-white in colour; claws blackish.

The female is rather larger than the male, and is said by some to be somewhat duller in color-ation; however, individuals vary in this respect. The bird is readily distinguished from all the other members of the Owl family by the smallness of its head.

Situation and Locality.—On the ground, amongst heather, long grass, rushes, sedge, and gorse; on large moors, upland heaths, fens, and marshes in Norfolk, Suffolk, Cambridgeshire, and in the northern counties of England and Scotland. It is said to be only a winter visitor to Ireland.

Materials.—Dry grass, moss, and other bits of dead vegetation, used sparingly to line the hollow made or selected; sometimes none whatever.

Eggs.—Three to five, generally; some authorities give the numbers as occasionally as many as seven or eight. White, and oval in form. Size about 1·6 by 1·28 in. Easily distinguished by nesting site.

Time.—April and May.

Remarks.—Resident and migratory, its numbers being swollen in winter by the arrival of more northern breeders. Notes, a shrill cry and snapping of the beak when the nest or young are in danger. Local and other names: Woodcock Owl, Hawk Owl, Mousehawk, Short-horned Howlet, Horned Oolert. A close sitter.

OWL, TAWNY. *Also* Wood Owl.

Description of Parent Birds.—Length about fifteen inches. Beak short, much curved, and horn white. Irides dark brown. The circle surrounding each eye is greyish-white, margined by a line of dark brown. Head, neck, back, and wings tawny-brown, finely marked with dark brown and black, and mixed with ash-grey. On the wing-coverts and scapulars are two descending lines of large white spots; the primaries are also barred with dark brown and dull white. Tail, two centre feathers uniform tawny brown, rest barred with tawny and dusky brown. Breast, belly, and under-parts greyish-white, streaked and mottled with two shades of brown. Under-coverts of tail white. Legs and toes covered with greyish-white feathers; claws large, much hooked, and horn white, with black tips.

The adult female is similar in plumage, but somewhat larger in size.

Situation and Locality.—The favourite nesting site is in a hole of a hollow tree, although the bird sometimes uses clefts of rock, holes in the walls of stables and barns, deserted nests of Rooks, Magpies, Crows, and Hawks; also rabbit-burrows. This Owl is a lover of woods, forests, and parks, and is pretty generally scattered over England, Wales, and the South of Scotland; rarer in the north, and almost absent from Ireland. I have met with it most numerously in Cumberland.

Materials.—None, the eggs being laid on decayed wood or the bird's own "castings."

Eggs.—Three to four. Pure white, smooth, and round. Size about 1·8 by 1·52 in. Distinguishing features, the round shape and large size.

Time.—March, April, May, June, July, August, September, October, and even later.

Remarks.—Resident. Notes : *tu-whit to-whoo;* when pleased the bird utters a low kind of whistle, and when angered snaps its beak with considerable sound. Local and other names : Brown Owl, Wood Owl, Jinny Oolert, Hoot Owl, Jenny Howlet, Ivy Owl. Comes forth at night and hoots weirdly. This bird is said to stand the light of day worse than any other member of the Owl family, although when fishing in rocky ravines I have seen it abroad on dull days at noon. A close sitter.

OWL, WOOD. *See* Owl, Tawny.

OYSTER-CATCHER.

Description of Parent Birds.—Length about sixteen inches. Bill long, straight, and orange-coloured. It is shaped like a vertical wedge, a form which renders it eminently useful for dislodging limpets and other bivalves from rocks. Irides crimson. Head, neck, back, and wings black, with the exception of a white, broad, slanting bar across the last. Rump and upper half of tail white, lower half black. There is a small patch of white under the eye. Throat and upper part of breast black. Lower breast and all the under parts of the body white. Legs and toes purple ; claws black. In the early spring the bird wears a white collar or gorget on the neck, but this disappears as the season advances.

Situation and Locality.—On the ground, amongst the shingle and sand of the sea-shore. Our illustration is from a photograph taken on the Farne Islands. Pretty generally in suitable localities round our coasts, and sometimes found quite inland on the banks of rivers and lakes. It is most numerous in Scotland and the surrounding isles.

Materials.—None, in the strict sense of the term, although a few shells or pebbles are often used as a kind of pavement, according to my experience. Sometimes a few bents are employed, or the eggs are laid on drifted seaweed.

Eggs.—Two to four, usually three. Yellowish, stone, or cream colour, streaked, blotched, and spotted with dark brown and grey. Occasionally the markings are inclined to form a zone at the larger end, but generally they are pretty evenly distributed over the shell. Size about 2·2 by 1·5 in.

Time.—May and June.

Remarks.—Resident. Notes, a clamorous chattering when the nest is approached. Local and other names: Sea Pie, Olive, Mussel Picker, Pienet, Tirma, Sea Piet, Trillichan, Chaldrick, Scolder, Sheldraker or Skelderdrake. Sits lightly, and generally has intimation of the approach of an intruder given by the male.

PARTRIDGE, COMMON.

Description of Parent Birds.—Length about twelve and a half or thirteen inches. Beak short, curved downwards, and bluish-grey. Irides hazel. Forehead and cheeks bright rust-colour; behind the eye is a patch of naked red skin; crown, back of the neck, and shoulders, cinereous-brown; lower back and wing-coverts mottled with two shades of reddish-brown on a pale buff ground, the central line of each feather being pale buff, unmarked.

OYSTER CATCHER.

Wing-quills brown, barred with pale buff. Tail-coverts and two centre quills brown, barred with rusty-red, remaining tail-feathers rusty-red; chin and throat bright rust-colour; neck, breast, and sides bluish-grey, freckled with a darker tinge of the same colour, marked on the breast with a horseshoe of rich bay, and barred on the sides with the same colour; under tail-coverts pale rusty-brown. Legs and toes bluish-grey; claws black.

The female is a trifle smaller than the male, and the rust-colour on her head is neither so extensive nor so bright, and the lesser and medium wing coverts and scapulars are marked with buff cross-bars, a feminine distinction first observed by Mr. Ogilvie Grant.

Situation and Locality.—On the ground at the bottom of a hedge (as in our illustration), amongst mowing grass, clover, standing corn, weeds, brackens, rough grass, and brambles. In ploughed fields, pasture lands, on the outskirts of woods, and in grass fields. Plentiful in cultivated districts, where preserved, but less numerous in high moorland districts, from which I have known the bird banished for years together by an exceptionally hard winter. In all suitable districts throughout the British Isles. Sometimes curious sites are chosen by this bird for its nest, such as on the thatch of a shed; and Booth mentions finding a Linnet's nest in the side of a stack, and that of a Partridge on the thatch of another close to it. I see, however, from my notes, compiled from the *Field*, that the latter is by no means an uncommon situation.

Materials.—A few blades of dry grass, bits of bracken or dead leaves, used as a lining to the slight hollow selected.

Eggs.—Ten to sixteen or twenty; as many

COMMON PARTRIDGE.

even as thirty-three have been recorded, but such
a large number is undoubtedly the production of
two females. Pale olive-brown or greenish-yellow,
unspotted. Size about 1·4 by 1·1 in. Easily distinguished from those of the Red-Legged Partridge
by smaller size, colour, and lack of spots.

Time.—May and June, although nests with
fourteen eggs in have been found as early as
April 18; and I have seen sitting hens have their
heads shaven off in grass-fields by the mower's
scythe as late as the middle of July.

Remarks.—Resident. Notes, *turwit* (call); *ajick*,
jick (alarm). Local and other names : none. Sits
very closely, and is of very uneven temper. Some
individuals will suffer a great amount of intrusion,
and others will forsake their eggs upon the slightest
molestation.

PARTRIDGE, RED-LEGGED.

Description of Parent Birds.—Length about
thirteen inches and a half. Bill short, curved
downwards, and red. Irides red. Crown bright
chestnut ; back of neck, back. rump, wing and tail-
coverts, brownish ; wing-quills darker and tipped
with light yellowish - brown ; tail - quills chestnut
and greyish-brown. A black streak runs from the
nostrils to the eyes. then turns downwards, making
a collar of black from which spots and streaks of
the same colour extend towards the upper part of
the breast. Breast pearly grey ; belly, vent, and
under tail-coverts fawn colour ; sides and flanks
transversely variegated with crescent-shaped marks
of black, white, pearly-grey, and fawn colour.
Legs, toes, and claws brown.

The female is not so large or bright and

RED-LEGGED PARTRIDGE'S NEST, WITH TWO
PHEASANT'S EGGS IN IT.

distinctive in coloration, and lacks the rounded
knob which takes the place of a spur on the leg
of the male.

Situation and Locality.—On the ground at the
bottom of hedgerows, amongst tall grass and other
herbage in corn, clover, and grass fields; occasionally it is said to select the thatch of a hayrick.
In cultivated and uncultivated districts, such as
commons and waste lands and heaths, more or
less in all parts of England, but most plentifully
in the southern and midland counties. Our illustration was obtained in Norfolk, and the nest
contained two Pheasant's eggs.

Materials.—Dry grass and dead leaves, used as
a lining to the hollow selected.

Eggs.—Ten to eighteen; yellowish-brown or
creamy-grey in ground-colour, spotted and speckled
with reddish or cinnamon-brown. The spots vary
in size and number, and the shell is coarse, pitted,
and very strong. Average size about 1·55 by 1·2 in.

Time.—April, May, and June.

Remarks.—Resident. It was introduced into
this country about two hundred years ago, but
has never gained a footing either in Scotland or
Ireland. Notes said to resemble *cockileke*. Local
and other names: Frenchman, French Partridge,
Guernsey Partridge. A close sitter.

PEEWIT. *See* LAPWING.

PETREL, FULMAR.

Description of Parent Birds.—Length about
nineteen inches. Beak of medium length, large,
strong, nearly straight, with exception of the upper

mandible, which is much curved downwards near the tip. It is of a yellowish colour, with a greenish tinge round the nostrils. Irides dark brown. Head and the whole of the neck white; back and wings French grey, except quills, which are darker; upper tail-coverts and tail-quills French grey; breast, belly, and under-parts white. Legs, toes, and webs pale grey. Many specimens are of an ash-grey or ash-brown tint all over, somewhat darker on the back and wings.

The female is similar to the male.

Situation and Locality.—Generally a kind of slight grotto or short burrow, often insufficient to hide the bird, dug by the Fulmar on turf-covered shelves and ledges; also in crevices of high, inaccessible rocks at St. Kilda, where there is a large colony. The bird has also established itself in the Shetland Islands within the last sixteen years.

Materials.—Dried grass and tufts of sea pink, sometimes nothing at all.

Egg.—One; white and rough when newly laid, but quickly becoming soiled. Average size about 2·9 by 1·98 in.

Time.—May and June.

Remarks. — Resident, but wandering. Note, Seebohm says it is "a very silent bird," and Macgillivray, "I never observed them" (the birds at St. Kilda) "utter any cry when flying, or even when their nests were being robbed." Local and other names: Fulmar, Northern Fulmar, Mallemock, Mallduck, Malmock. Gregarious, and a close sitter.

PETREL, LEACH'S FORK-TAILED.

Description of Parent Birds. — Length about seven and a quarter inches. Bill of medium length, nearly straight, and black. Irides dark brown. Head, neck, and back brownish-black, the two former rather lighter in shade than the latter. Wing-coverts reddish-brown, tinged with grey on the edges; quills black. Upper tail-coverts white; tail-quills black and slightly forked; breast, belly, vent, and under tail-coverts, in the middle, black. A white streak starts from behind each thigh and runs down the sides of the vent and under tail-coverts. Legs, toes, webs, and claws black.

The female is like the male.

Situation and Locality. — In burrows made in soft peat earth, under rocks, holes, fissures, and clefts in rocks, and in holes of walls, close to the sea; on rocky islands, such as the St. Kilda group, Hebrides, and some of those off the Irish Coast.

Materials. — Dry grass and bits of moss, sometimes nothing whatever.

Egg. — One; white, chalky, and speckled round the larger end with small rust-coloured and greyish-brown spots. Size about 1·3 by ·96 in.

Time. — June.

Remarks. — Resident, but wandering. Notes: *peer-wit, peer-wit*, said to be uttered by the birds as they sit on their nests, also both night and day. Local and other names: Leach's Petrel, Fork-tailed Storm Petrel, Fork-tailed Petrel. Gregarious, and a very close sitter.

PETREL, STORM.

Description of Parent Birds.—Length about six inches. Bill moderately long, hooked at the tip, and black. Irides dark brown. Head, neck, back, wings, and tail a uniform sooty black. The outer edges of some of the smaller feathers of the wings and upper tail-coverts white. All the under-parts are sooty brown, with exception of the sides of the vent, which are white. Legs, toes, and webs black.

The female does not differ from the male.

Situation and Locality.—In old Puffin and rabbit-burrows, holes in cliffs, under large boulders, and in holes in walls. In the Scilly Islands, Lundy, at suitable places along the Welsh coast, the western and northern coasts of Scotland, and the islands lying off them; round the Irish coast, but neither on the east coast of England nor Scotland. Our illustration represents a boulder of rock under which a Stormy Petrel had its nest on Ailsa Craig for several years, according to Cragsman Girvan, who lives upon the rock.

Materials.—A few blades of dry grass generally, but the egg is often laid on the bare ground.

Egg. — One. White, rough, and chalky in appearance, with small, dust-like reddish-brown spots in an almost indistinguishable zone round the larger end. Size about 1·1 by ·83 in.

Time.—End of May, June, July, and even as late as September.

Remarks.—Resident, but keeping to the open sea, except during the breeding season, or when driven ashore by violent storms. Notes, a warbling chatter, sung whilst the bird is sitting on her

N

egg. Local and other names: Mother Carey's Chicken, Stormy Petrel, Little Petrel, Witch, Allamotti, Sea Swallow, Spency, Assilag, Mitty. Sits very closely.

PHALAROPE, RED-NECKED.

Description of Parent Birds.—Length from seven to eight inches. Bill of medium length, straight, slender, and black. Irides dark brown. Head, hind part of neck, back, wing-coverts, scapulars, and tertials dark ash-grey, the last two sets of feathers being tipped with rust-colour. Wing-quills dusky, some of them being tipped with white. Rump and upper tail-coverts dusky, banded with white : tail-quills dusky or brownish-grey, darkest in the centre. Chin white ; front and sides of neck rusty red ; upper breast grey, barred with white; under-parts white. Some specimens are white from chin to vent. Legs, toes, and membrane down either side of toes, green ; claws black.

The female is rather larger, and more richly coloured.

Situation and Locality.—On the ground, in tufts of grass, in a hollow on the top of a small hillock; on moors and mountains not far from the edge of a loch or pool, in the Orkneys, Shetlands, and Hebrides only, according to Mr. Dixon, but according to Mr. Saunders (writing, I ought to mention, ten years earlier), in some parts of the mainland. The latter author was, at the time of writing, doubtful whether the bird had been extirpated in the Orkneys. However, I have evidence of its nesting there as late as 1892.

As some evidence of how the bird is gradually being banished from our shores, Mr. Arthur Orde

SITUATION OF STORM PETRELS NEST

+ Entrance-hole.

told me, whilst in North Uist, that its eggs fetch
as much as ten shillings each in that island.

Materials.—Dry grass. The nest is said to be
deep, and about the size of that of a Titlark.

Eggs.—Four, ground colour varying from olive-
green to light buffish-brown, spotted and blotched
with umber and blackish-brown, most thickly at the
larger end. Size about 1·1 by ·83 in. Easily distinguished by small size.

Time.—June.

Remarks.—Migratory, arriving in May and departing in August. Notes, *tirrr.* Local and other
names: Red Phalarope, Half Web, Red-necked
Coot Foot, Red-necked Lobe Foot. Gregarious,
and very tame on its breeding-grounds.

PHEASANT.

Description of Parent Birds.—Length about three
feet, nearly two of which are accounted for by
the abnormally long tail. Beak short, curved
downwards, and light yellowish horn colour, duller
at the base. Irides hazel. Round the eye the
skin is bare, crimson, and minutely speckled with
black. The feathers of the head and neck all
round are steel-blue, with a purple-green or brown
sheen, according to the light upon them. Upper
part of back deep orange, tipped with rich black;
lower back and scapulars made up of a mixture of
dark orange and dull brown, each feather having a
straw-coloured outer margin. Wing-coverts red, of
different shades. Quills greyish and yellowish-brown.
Rump and upper tail-coverts pale yellowish-brown.
Tail-quills brown, tinged with yellow, and transversely barred with black. Breast and belly golden

PHEASANT.

red, each feather edged with rich black, glossed with
purple and gold; vent and under tail-coverts dusky
brown. Legs, toes, claws, and spurs brownish-grey.

The female is about a foot shorter, and much
subdued in coloration. Her plumage is composed
principally of yellowish and dark brown.

Situation and Locality.—On the ground, amongst
coarse, long grass, in or near hedgerow bottoms,
under bramble bushes, brackens, weeds, and scrub,
on the outskirts of woods, plantations, and coppices
all over the country where there is plenty of wood,
water, and protection. Specimens have been found
occupying a deserted squirrel's drey, in a Scotch
fir, and on the tops of stacks. Our illustration
represents one amongst tall dead grass at the foot
of a hedge.

Materials.—A few dead leaves, dried grass-blades,
bracken, or fern-fronds.

Eggs.—Eight to thirteen; as many, however,
as seventeen have been found in one nest under
circumstances which pointed to their having been
laid by one hen, and in other cases even as many
as twenty-six, undoubtedly the joint production of
two hens, as the bird has often been known to share
a nest, not only so far as laying was concerned,
but sitting with other hens of its own species;
also with the Partridge, and has been known to
lay both in the nest of the Red Grouse and
Capercaillie. Olive-brown is the general colour of
the eggs, but specimens may be met with of a
greyish-white, tinged with green or bluish-green.
They are unspotted, but finely pitted. Size about
1·87 by 1·4 in.

Time.—April and May, sometimes as early as
March, and as late as September or October.

Remarks.—Naturalised, and holds its own only

by protection. It is thought that the bird was first introduced by the Romans and brought from the neighbourhood of the Black Sea. Notes: crow of male, a short, loud cackle, and the note of the female, a shrill, piping whistle. Local and other names, none. Sits close and, curiously enough, emits no scent at this period. In a wild state the bird is monogamous, but in this country, in its semi-domesticated condition, it is polygamous.

PINTAIL DUCK. *See* DUCK, PINTAIL.

PIPIT, MEADOW.

Description of Parent Birds.—Length about six inches. Bill of medium length, slender, straight, and dark brown, except at the base of the under-mandible, where it is of a lighter tint. Irides hazel. Crown, nape, back, and upper tail-coverts dark brown, the border of each feather being of a lighter greyish tint. Wings brownish-black, the feathers being edged with light brown; tail dark brown, the two outer feathers on either side margined with white, the rest with light brown; chin and throat dull white; sides of neck and breast pale buffish-white or yellowish-brown, with numerous elongated dusky spots; belly and under tail-coverts dull white, tinged with brown. Legs and toes light brown; claws dusky, hind one long and curved.

The female is said to be slightly smaller, though the difference is not at all apparent; her plumage is similar.

Situation and Locality.—On the ground in the shelter of a tuft of grass, heather, bit of overhanging bank or stone. I was shown one in a hole several inches deep in a bank from which the earth had all slipped away last summer. The bird nests commonly throughout the British Isles, but most numerously in pasture land and moorland districts. Our illustration was procured on the Westmoreland hills.

Materials.—Bents, bits of fine dead grass and horsehair.

Eggs.—Four to six, generally five ; French grey, sometimes tinged with pale bluish-green, thickly covered with light or dusky brown. The markings are generally so thickly distributed as to hide the ground-colour ; indeed, I have met with specimens where none of it could be seen. Occasionally eggs may be found marked with hair-lines of dusky black at the larger end. Size about ·8 by ·58 in. Distinguished by small size and brown appearance.

Time.—April, May, June, and occasionally as late as July.

Remarks.—Migratory and resident, the latter being subject to local movement. Notes: song, short, soft, and musical ; alarm notes, *trit, trit ;* call, *zeeah, zeeah, zeeah.* Local and other names : Titling, Moor Tite, Titlark, Ling Bird, Teetick, Moss Cheeper, Wekeen, Pipit Lark, Heather Lintie, Moor Titling, Moor Tit, Meadow Lark. Sits closely, and hovers round, uttering its note, *trit, trit.*

PIPIT, ROCK.

Description of Parent Birds.—Length about six and three quarter inches. Bill medium, nearly straight, slender, and dark brown, except at the base, where it is dull orange. Irides dark brown.

MEADOW PIPIT.

A faint yellowish-white streak runs over the eye and ear-coverts. Crown, nape, back, rump, upper tail-coverts, and wings dull brown, slightly tinged with green, the feathers being streaked along the centre with a darker shade. Edges and tips of wing-quills lighter. Tail dark brown; outer feathers on either side greyish, on the exterior webs and at the tips. Chin and throat greyish, or dull yellowish-white, the latter and sides of neck mottled and streaked with brown. Breast dull greenish-white, streaked and spotted with brown. Sides olive-brown; belly, vent, and under tail-coverts dull yellowish-white, sparingly streaked with brown. Legs and toes reddish-brown; claws black.

The female is a trifle smaller, but similar in plumage.

Situation and Locality.—On ledges and in crevices of rock; under an overhanging piece of stone, or in the shelter of a tuft of grass growing on rocky sea coasts, pretty generally round our shores, with the exception of Essex, Suffolk, Norfolk, and Lincoln. Our illustration was obtained on the Farne Islands.

Materials.—Seaweed and dry grass of various kinds, with an inner lining of finer grass, and occasionally horsehair.

Eggs.—Four to five, grey in ground colour, slightly tinged with green or reddish-brown, minutely and closely spotted and mottled with underlying markings of grey, and surface spots of reddish-brown, occasionally marked at the larger end with one or two dark brown lines. The spots are, as a rule, more numerous at the larger end. Size about ·85 by ·63 in. Distinguished by large size and locality of the nest.

Time.—April, May, June, and July.

ROCK PIPIT.

Remarks.—Resident, although numbers migrate. Notes: call, a shrill *hist* or *pst.* Local and other names: Shore Pipit, Rock Lark, Sea Titling, Dusky Lark, Field Lark, Sea Lintie. Sits close.

PIPIT, TREE.

Description of Parent Birds.—Length about six and a half inches. Bill of medium length, nearly straight, slender, and dark brown, lighter on the edges and at the base. Irides hazel. Crown, nape, and back dark brown, the feathers being bordered with lighter brown; wings darkish-brown; lesser coverts edged and tipped with greyish-white; greater coverts edged with pale brown; these two lighter colours form distinct bars across the wings. Rump, upper tail-covers, and quills brown, the two outer ones on each side nearly all dirty white; chin and throat pale brownish-white. A brown streak runs from the gape, slightly backward, and for some distance downwards. Sides of neck and breast pale buff, with streaks of brown on the former and round spots on the latter; belly, vent, and under tail-coverts dirty white. Legs, toes, and claws pale yellowish-brown.

The female is a little smaller in size, and the spots on her breast not so large. It is larger than the Meadow Pipit, has a stronger beak, fewer and larger spots on the breast, and the claw on the hind toe is shorter.

Situation and Locality.—On the ground, concealed by a tuft of grass, in hedgerow banks, on the sloping banks of streams, hidden by a low weed-tangled bush. The bird seems fond of the same locality, and I know several places in

TREE PIPIT.

Yorkshire where pairs return to nest year after year
with the utmost regularity; the cocks using the
same tree, often the very same branch, to start
from and return to after their short singing flight.
Near woods, plantations, and tree-fringed streams.
Affects more cultivated districts than the Meadow
Pipit. Scattered over England in suitable districts,
rare in the west and Wales, more numerous in
the south of Scotland, rare in the north, and not
reported in Ireland, on reliable authority. Our
illustration was procured in North Yorkshire.

Materials.—Dry grass, moss, roots, lined with
finer grass and generally, though not always,
horsehair.

Eggs.—Four to five, sometimes six; exceedingly
variable in coloration. Professor Newton regards
those of "a french-white, so closely mottled or
speckled with deep brown as almost to hide the
ground colour," as the normal type; whilst Morris
regards those "greyish-white in ground-colour, with
a faint tinge of purple, clouded and spotted with
purple-brown or purple-red," as the most general,
and amongst the nests I have found this type has
certainly been the most numerous. In another
type the ground colour is yellowish-white, and the
spots rich reddish-brown. Some eggs are of a
uniform brownish-pink, rarely marked on the larger
end with hair-lines of dark brown or black. The
spots not only vary much in colour but in size
and distribution. Size about ·83 by ·63 in. The
larger size of the eggs, their inclination to reddish-
brown, and locality of nest, help to distinguish them.

Time.—May and June.

Remarks.— Migratory, arriving in April and
departing in September or October. Notes, a sweet,
ringing *tsee*, *tsee*, *tsee*, uttered pretty quickly.

Local and other names : Field Lark, Field Titling, Pipit Lark, Tree Lark, Grasshopper Lark. Sits closely, especially when incubation is advancing.

PLOVER, GOLDEN.

Description of Parent Birds.—Length between eleven and twelve inches. Bill of medium length, straight, and black. Irides brownish-black. On the forehead is a band of white ; crown, nape, neck, back, wing-coverts, rump, and tail-coverts, deep greyish-brown, the edges and tips of the feathers being marked with yellow ; wing-quills brownish-black ; tail dark brown, barred with brownish-black and greyish-white. Chin, throat, sides of neck, breast, and belly deep rich black, bordered on the sides below the wings by a band of white ; under tail-coverts white. Legs, toes, and claws black.

The female resembles the male, but both are subject to some little variation, depending upon constitutional vigour.

Situation and Locality.—On the ground. I have met with nests in short closely-cropped heather that were nearly as cup-shaped as the nest of the Chaffinch, and in which the four eggs were almost standing on their sharp ends. Amongst fringe moss, coarse grass, and short heather in the rough moorland and wild boggy parts of Somerset and Devon, the North of England, Wales, Scotland, and Ireland.

Materials.—A few pieces of dry grass, rushes, or heather-tops, forming a lining to the hollow of the nest.

Eggs.—Four, pear-shaped, yellowish stone or cream colour, blotched and spotted with umber-brown and blackish-brown. They are larger than those of the Lapwing, not quite so pyriform, and lack the olive tinge in their ground colour. The birds also nest, as a rule, on higher and wilder ground. Size about 2·07 by 1·4 in.

Time.—May and June.

Remarks.—Resident, but subject to local and partial migration. Notes, *tlui*, and *taludl, taludl, taludl*, the first note being uttered with low and melancholy deliberation and the latter hurriedly. Local and other names: Yellow Plover, Whistling Plover. Does not sit very close as a rule; however, I have known the bird do so before incubation was far advanced, and feign a broken wing in order to decoy the intruder away.

PLOVER, GREAT. *See* CURLEW, STONE.

PLOVER, KENTISH.

Description of Parent Birds.—Length about seven inches. Bill shortish, nearly straight, and black. Irides brown. Forehead and a line running over the eye and ear-coverts white; middle crown black; back of head yellowish-brown. A black streak commences at the base of the beak, and passing through the eye, includes the ear-coverts; nape white; back, wings, and upper tail-coverts ash-brown, with exception of the wing-primaries, which are dull black, edged with white on some of the outside shafts; tail-quills ash-brown towards the base, dusky black towards the tip, and white on the outsides; chin, cheeks, and sides of upper part of neck white; sides of lower part of neck, just in front of the shoulder or point of the wing, black; breast, belly, and under-parts white. Legs, toes, and claws dark slate-colour.

The female differs only in having the black on the head and sides of lower neck less distinct and covering a smaller area.

Situation and Locality.—In a hollow of the sand or shingle; sometimes on dry seaweed which has been cast up by the waves. The breeding area of the bird is very limited indeed, and its numbers are gradually decreasing. In suitable places along the coast between Hastings and Dover.

Materials.—None, the eggs being deposited in a slight hollow.

Eggs.—Three to four, generally the former number; cream, stone, or dark buff in ground colour, streaked, spotted, and blotched with brownish-black and very dark grey. Size about 1·25 by ·9 in. More pyriform than those of the Lesser Tern, and distinguished from those of the Ringed Plover by scrawl-like character of markings.

Time.—May.

Remarks.—Migratory, arriving in April or early in May, and departing in August or the beginning of September. Notes: call, *tirr, tirr, pitt, pitt, pwee, pwee;* alarm, a plaintive and also a sharp whistle. Local and other names: none. The bird is gradually becoming scarcer as a breeding visitor, and will probably, as such, become extinct before long. In Yarrell's time the eggs were in great demand as table delicacies, and dogs were trained to find them. Does not sit closely, and runs on quitting the nest.

PLOVER, NORFOLK. *See* CURLEW, STONE.

O

PLOVER, RINGED. *Also* Ringed Dotterel.

Description of Parent Birds.—Length about seven and three-quarter inches. Beak short, straight, black at the tip, and rich, dark yellow towards the base. Irides brown. Forehead white, middle of crown black, followed by greyish-brown, which extends down the back of the neck; back and wings greyish-brown, except the ends of the coverts, which are tipped with white, and the primaries, which are dusky, with some white at the base and along the shafts. Upper tail-coverts greyish-brown; quills greyish-black in the centre, outside feathers white. A black patch commences at the gape and passes under the eye, backward and downward, to the side of the neck. A broadish collar of white passes round the upper part of the neck, followed by a gorget of black, which is deepest in front. Breast and all the under-parts white. Legs and toes orange; claws black.

In the female the black parts of the head and neck are not so broad or well defined.

Situation and Locality.—On the plain surface or in a slight hollow, scraped in the sand or shingle above high water-mark on stretches of flat, sandy shores, also in shallow crevices of bare, flat sea-shore rocks; sometimes quite inland on the banks of rivers and lakes in nearly all suitable places throughout the British Isles. Our illustration was procured near to Bamborough Castle.

Materials.—None generally, but sometimes a lining of small pebbles; and in places where a crevice in a flat rock has been adopted I have often met with a lining of small sea shells.

Eggs.—Four, pear-shaped, pale buff, cream, or

RINGED PLOVER.

stone colour, spotted with smallish and pretty evenly distributed dots of black, blackish-brown, and bluish-grey, which distinguish them from those of the Kentish Plover. Size about 1·4 by 1·0 in.

Time.—April, May, and June, although eggs have been found as early as March and as late as August.

Remarks.—Resident, but subject to much local movement. Notes: alarm, *pen-y-et.* Local and other names: Sand Lark, Dull Willy, Sand Lavrock, Ringed Dotterel, Stonehatch. Does not sit closely. Somewhat gregarious. Eggs harmonise very closely with their surroundings.

POCHARD.

Description of Parent Birds.—Length about nineteen inches. Bill of medium length, depressed near the centre, black at the tip, pale blue in the middle, and black at the base, which is slightly raised. Irides red. Head and upper part of neck, all round, deep chestnut, lower part of neck, all round, black. Back, scapulars, tertials, and wing-coverts finely freckled with grey and dusky undulating lines; secondaries and primaries bluish-grey, the former tipped with white and the latter ending in dusky brown. Rump and upper tail-coverts dusky black; tail feathers dusky brown mixed with grey. Breast, sides, and under-parts greyish-white, marked by minute dusky lines, darkest on the vent. Legs, toes, and webs leaden grey.

The female differs considerably in appearance. Her bill is black. Irides reddish-brown. Head and upper part of neck dull greyish-brown, lightest in front; lower portion of neck dusky brown. Upper-

POCHARD.

parts darker, and under-parts dull greyish-white, clouded with brown; under tail-coverts dusky grey.

Situation and Locality.—On the ground, in tufts of rushes, coarse grass, in osier beds; amongst flags and sedges growing on the shores of lakes, broads, and tarns, in the North, East, and South of England; also in Scotland and Ireland. Our illustration was procured in Norfolk. The nest was situated amongst the reeds on the right-hand side of the picture opposite.

Materials.—Sedges, rushes, and dry grass, with an inner lining of down. The tufts are brownish-grey with whitish centres.

Eggs.—Seven to ten, occasionally as many as thirteen or fourteen. Pale greyish-buff or greenish-drab. Size about 2·35 by 1·7 in. Distinguished from the Tufted Duck only by the browner colour of the down-tufts.

Time.—May.

Remarks.—Resident and migratory, being more numerous in winter. Notes, a low whistle, but when alarmed or vexed, a hoarse kind of croak, like *kr-kr-kr.* Local and other names: Red-headed Wigeon, Duncur, Red-headed Poker, Dunbird, Vare-headed Wigeon, Attile Duck, Blue Poker, Great-headed Wigeon. Sits closely.

PTARMIGAN.

Description of Parent Birds.—Length fifteen inches. Bill short, strong, curved downwards, and black. Irides hazel. Over the eye is a piece of erectile skin of a bright red colour. Head and neck barred and mottled with black, rusty brown, and white or grey; back and upper tail-coverts

TYPICAL NORFOLK DUCK MERE.

pale brown or ash, mottled with small dusky spots
and bars. Wings white, the shafts of the quills
being black. Tail-quills black, tipped with white,
the two centre feathers sometimes grey; chin white;
throat white, mottled with brown; breast same
as back; belly, vent, and under tail-coverts white.
Legs and feet dull white; claws black.

The female is a trifle smaller; her head and
upper-parts have more red, rusty yellow, and black,
and less grey than in the case of the male. The
dark parts on the wing-quills are broader, and her
under-parts are darker.

Situation and Locality.—On the ground, amongst
heather and the vegetation growing on the rock-
strewn and bleak mountains of the Highlands of
Scotland, and some of the larger islands of the
Hebrides.

Materials.—A few bits of dead heather, dry
grass, or leaves, used as a lining to the hollow
chosen for the reception of the eggs.

Eggs.—Seven to ten or twelve, greyish-white
to pale red-brown in ground colour, blotched and
spotted all over with very dark, rich brown. Size
about 1·7 by 1·1 in. Distinguished from the eggs
of the Red Grouse by their buffy ground colour
and smaller number of markings.

Time.—May and June.

Remarks.—Resident. Notes, sometimes low, and
at others a kind of loud and prolonged croak.
Local and other names: Rock Grouse, White
Grouse, White Partridge (from the fact that the
bird turns white in winter), White Game. Sits
very close, and the nest is difficult to find.

PUFFIN.

Description of Parent Birds.—Length about twelve inches. Bill rather short, and deeper than it is long. Both mandibles are arched from base to tip, the upper one being a trifle hooked. It is of such clumsy appearance as to suggest a kind of sheath over the real bill. It is much compressed sideways, and furrowed transversely. The basal ridge is yellow; then occurs a space of bluish-grey, followed by four ridges and three grooves of a rich orange colour. There is a space of naked skin at the gape, which is yellow. Irides grey. Cheeks and ear-coverts dirty white; forehead, crown, back of head, ring round neck, back, wings, and tail, black. Breast, belly, and vent white. Legs, toes, and webs orange; claws black.

The female has a slightly narrower bill.

Situation and Locality.—In burrows of varying length, dug by the bird's own exertions, in peat or mould, or taken from a rabbit by force; sometimes amongst fallen rocks or in crevices of cliffs. Our illustration is from a photograph taken on the Farne Islands, where a large colony nests yearly. I took the egg from the end of one of two burrows having the same entrance, and placed it in front so as to show in the picture. In walking across the top of the island, which is covered by a soft layer of peat, the visitor feels the earth giving way beneath his feet at each step, so much is it honeycombed by the birds, which scuttle out of their burrows in all directions. Breeds at the Farne Islands, Flamborough Head, parts of the south coast, Scilly Isles, west coast of England, Wales, Lundy, west coast of Scotland, Hebrides,

North of Scotland, and in suitable places round the Irish coast.

Materials.—Occasionally a few bits of grass, feathers, or roots; often nothing whatever.

Egg.—One; dull white or grey, marked with a few indistinctly defined spots of pale brown and grey, generally at the larger end. They soon become soiled and dirty from contact with their surroundings. Size about 2·4 by 1·67 in.

Time.—May and June.

Remarks.—Migratory, arriving in April and departing in August. More northern breeding birds winter with us. Notes, o-r-r to a-r-r, according to the bird's state of mind. Local and other names: Tammy Norrie, Coulter-Neb, Sea Parrot, Tommy Tonnoddy, Ailsa-Cock, Cockandy, Lunda Bonger, Gulderhead, Bottlenose, Pope, Marrot, Mullet. Sits closely.

QUAIL.

Description of Parent Birds.—Length about seven inches. Beak rather short, strong, slightly curved, and dusky. Irides hazel. A broadish, pale wood-brown streak commences at the base of the beak, and passes over the eye and ear-coverts. The crown is also divided by a much narrower streak of the same colour. The feathers of the head, neck, back, rump, and tail are brown, with lighter coloured shafts and longitudinal streaks of wood-brown. Wing-quills dusky brown, with small rust-coloured bands on the outer webs. Chin and throat white, crossed by two dark brown gorgets; breast pale rusty brown with lighter shafts; under-parts yellowish-white; flank-feathers pale buff in

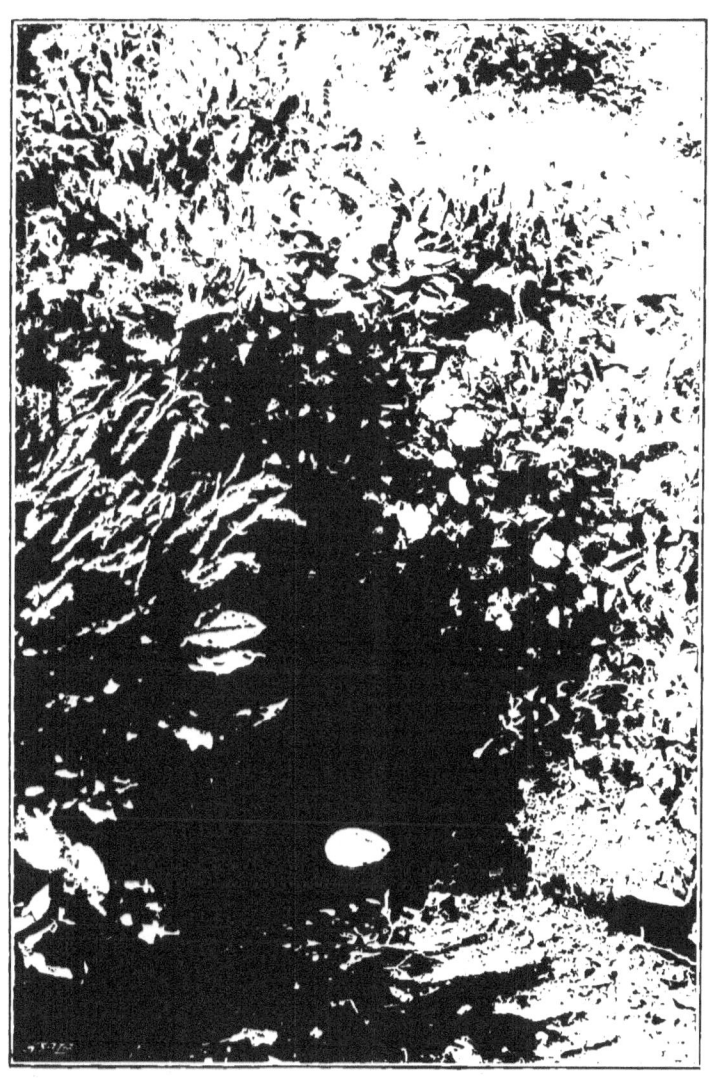

PUFFIN.

the centre, mottled and barred on the margins with brown. Legs, toes, and claws pale brown.

The female lacks the gorgets, and has the breast feathers marked with dark spots on either side of the pale shafts.

Situation and Locality. — On the ground in growing cornfields, grass and clover fields, sparingly throughout the British Isles.

Materials.—The slight hollow used as a nest is scantily lined with blades of grass, trodden down, or a few dead leaves.

Eggs.—Seven to twelve ; as many even as twenty have been found, doubtless the production of two females. Pale yellowish-brown of varying shades, spotted, blotched, and clouded with umber-brown and blackish-brown. Some eggs are only spotted, whilst others are thickly clouded with varying shades of brown. Size about 1·1 by ·9 in. Distinguished by small size and character of markings.

Time.—May and June.

Remarks.—Migratory and resident. Notes, *click-clic-lic.* Other names : none. Sits closely.

—

RAIL, LAND. *See* CRAKE, CORN.

RAIL, WATER.

Description of Parent Birds.—Length nearly twelve inches. Beak rather long, nearly straight, and red. Irides hazel. Crown, neck, back, and wing-coverts olive or reddish-brown, with a deep black mark in the middle of each feather : wing and tail primaries dusky, with lighter margins ; chin, cheeks, throat, sides of neck, breast, and belly, leaden-grey; sides and flanks deep slaty-grey, with bars of white; vent buffish; under tail-coverts greyish-white. Legs and toes reddish-brown.

The female resembles the male, although her beak is not so long or her plumage so bright and distinctive; she also generally shows some white bars, which the male lacks, on the wing-coverts.

Situation and Locality.—On the ground amongst long grass, a clump of rushes or reeds, in thick osier beds, swamps where alders grow, round ponds and ditches, on the banks of slow-running rivers and in boggy ground abounding in reeds and dense aquatic growths; generally throughout the United Kingdom, but nowhere abundant. Most numerous in the eastern counties of England.

Materials.—Reeds, sedge grass, and flags, in rather liberal quantities.

Eggs.—Five to eleven, generally six or seven; creamy-white in ground colour, speckled with a few small reddish spots and underlying dots of ash-grey. Size about 1·4 by 1·0 in. The locality of the nest and small spots distinguish them from those of the Corn Crake.

Time.—April, May, June, and July.

Remarks.—Migratory and resident. Many of our winter visitors retire North to breed. Note, a soft *whit*, heard after dusk. Local and other names: Runner, Skiddycock, Brook-runner, Bilcock, Velvet-runner, Grey-skit, Oarcock. A pretty close sitter, slipping away without demonstration.

RAVEN.

Description of Parent Birds.—Length about twenty-six inches. Beak of medium length, curved

downward towards the tip, stout, and black. Irides
brown and grey. At the base of the beak are a
number of coarse hairs pointing forward. The
plumage is a uniform black, glossed with a purple
and blue sheen. Legs, toes, and claws black.

The female is a little smaller, and less glossed.

Situation and Locality.—In crevices and on
ledges of high inaccessible cliffs, either on the sea-
shore or inland. Our illustration is from a photograph taken in Mull. The nest was in an exceedingly difficult situation to photograph or approach
in any way, and consequently does not show up
very clearly in the picture : it is two and five-
eighth inches from the bottom, and one and three-
quarter inches from the right-hand side of the
picture. The bird sometimes nests in tall trees,
and is to be found in the wild unfrequented parts
of England, Wales, Scotland, and Ireland.

Materials.—Sticks of various sizes (I have seen
them made entirely of juniper), wool, and hair.

Eggs.—Five to seven ; greyish-green, bluish-
green, or greenish-brown, blotched, splashed, and
spotted with dark greenish or smoky brown, and
underlying markings of a lighter greyish-purple
tinge. Variable both in regard to coloration and
size, but generally distinguishable from those of
the Carrion Crow and Rook by their larger size.
Average measurement about 1·95 by 1·3 in.

Time.—February, March, and April.

Remarks.—Resident. Note, a deep hoarse *cronk*,
that may be heard at great distances. Local and
other names : Corbie, Corbie Crow, Great Corbie
Crow. Sits lightly.

RAVEN.

RAZORBILL.

Description of Parent Birds. — Length about seventeen or eighteen inches. Bill fairly long, straight, except towards the tip, where it is much decurved, and black. A white, curved line runs across both mandibles, and a well defined one from the top of the bill to the eyes. The basal half of the beak is covered with feathers. Irides dark brown. Crown, nape, back, wings, and tail black, with the exception of a narrow band of white across the wings; chin and throat dark brown; breast and all under-parts snowy white. Legs, toes, and webs brownish-black; claws black.

Female similar.

Situation and Locality. — In crevices, crannies, under crags, and on ledges of high maritime cliffs; pretty generally round our coasts One observer says that he has found its nest in a Puffin burrow, and another in a Cormorant's nest at the Farne Islands. The latter is a somewhat remarkable circumstance, inasmuch as the Cormorants occupy a rock exclusively there. Our illustration is from a photograph taken on Ailsa Craig, where great numbers breed.

Materials. — None; the egg, when laid on bare, flat rock, is often swept off by a gust of wind.

Egg. — One; varying from white to buffy-white, or even reddish-brown, spotted and blotched with large, bold and numerous markings of greyish, chestnut, reddish and blackish brown. Average size about 2·9 by 1·87 in. Not so pointed as that of the Guillemot, and interior of shell greenish instead of yellowish-white, which is the colour of all varieties of Guillemots' eggs, except those with intense green or blue ground colours, when blown.

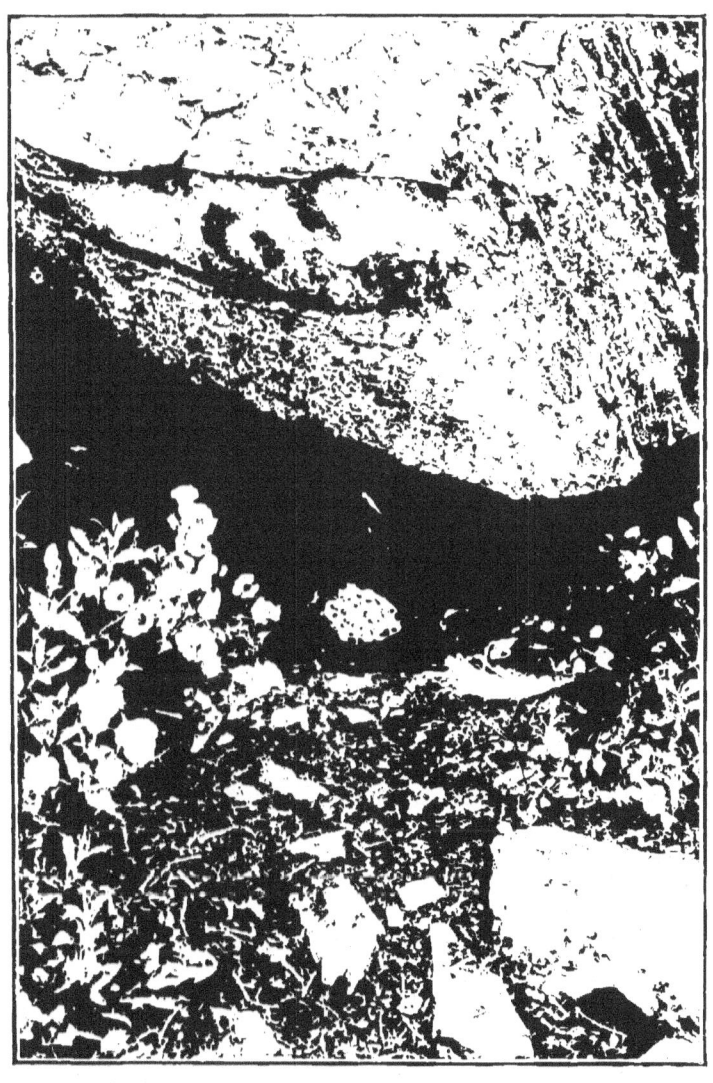

RAZORBILL.

P

Time.—May and June.

Remarks. — Resident, but wandering, except during the breeding season. Notes, a kind of grunting noise when disturbed. Local and other names : Black-billed Auk, Marrot, Murre, Razorbill Auk, Sea Crow, Bawkie, Alk or Oke, Falk ; in Zetland, Hiogga. Gregarious, and a close sitter.

REDBREAST. *See* Robin.

REDPOLL, LESSER.

Description of Parent Birds.—Length about four and a quarter inches. Bill short, nearly conical, thick at the base, and brownish horn colour. Irides dusky brown. Crown, to beyond the line of the eyes, crimson red ; sides of head brown. Hinder part of crown, nape, back, rump, and upper tail-coverts dark brown, the feathers being bordered with light reddish-brown, slightly mixed with grey ; tail-coverts tinged with crimson. Wings dusky, the feathers being edged with pale brown ; the middle and greater coverts are tipped with light reddish-brown, making two rather showy bars across. Tail slightly forked and dusky, edged outwardly with pale brown. Chin black, throat and breast rose-pink to vermilion ; the middle of the breast, flanks, and under-parts light greyish. Sides streaked with dull brown. Legs and toes darkish brown ; claws nearly black.

The female is a trifle smaller, and lacks the red on her breast and upper tail-coverts. Chin brownish-black ; under-parts brownish-white, streaked with dark brown on the breast, sides, and flanks.

Situation and Locality.—In alders, willows, elms, firs, hawthorn, hazel and other trees and bushes, generally pretty low down, but sometimes at a considerable height. Occasionally it may be found in a heather tuft. In shrubberies, coppices, plantations, and bushes that fringe streams and ponds in mountain districts. It has been found breeding in nearly every county in England, but is most numerous in the northern counties and in Scotland. It breeds in Ireland most numerously in the north.

Materials.—Fine twigs (used as a foundation), dry grass, stalks, moss, and roots, with an inner lining of willow down, sometimes hair and feathers. It is cup-shaped, and, as a rule, a well made and beautiful little structure.

Eggs.—Four to six, generally five, of a very pale bluish-green colour, spotted generally about the larger end with orange-red, occasionally streaked with a darker colour. There are also underlying markings of pale greyish-brown. Size about ·62 by ·46 in. The black chin and smaller size of the parent birds, and their eggs, distinguish the nest of this bird from that of the Linnet or Twite.

Time.—May and June.

Remarks.—Resident in Scotland and the North of England. A winter visitor further south, generally speaking, though specimens have stayed and bred. Notes: call, *peewit* and *kreck, kreck, hayid!* song meagre but lively. Local and other names: Lesser Redpoll Linnet, Lesser Red-headed Finch, Rose Linnet. Sits very closely indeed.

—

REDSHANK.

Description of Parent Birds.—Length about eleven inches. Bill long, straight, slender, and dusky at the point and reddish at the base. Irides hazel. Crown, nape, back and wing-coverts greyish-brown, spotted and streaked with black ; secondaries tipped with white ; primaries nearly black. Rump, tail-coverts, and feathers white, the last barred with dusky black. Over the eye is a white streak, and from the gape to the eye a dusky brown one. Chin, throat, breast, and under-surface of the body greyish-white, spotted and streaked with brownish-black. Legs and toes red ; claws black.

The female resembles the male, but is larger.

Situation and Locality.—On the ground, in a little hollow or depression, sheltered by an over-hanging tuft of coarse grass or heather, or in the crown of a rush-root, generally well concealed ; in fen, marsh, and boggy districts on the swampy shores of mountain tarns and lochs ; in the eastern counties of England ; also in suitable parts of Scotland and Ireland.

Materials.—A few blades of grass or bits of moss ; often nothing at all.

Eggs.—Four, much pointed at the smaller end ; ground colour varying from pale straw to buffish-brown, spotted and blotched with rich dark brown, and underlying markings of light brown and grey. Size about 1·78 by 1·23 in. Distinguished by the buff ground colour and bold blotches.

Time.—April and May.

Remarks.—Resident, but subject to local move-ment. Notes : alarm, a shrill, discordant cry, resembling *took* or *tolk*. Local and other names : Redshank Sandpiper, Pool Snipe, Red-legged Horse-man, Sandcock, Red-legged Sandpiper, Teuke. Sits lightly, and when incubation has advanced, resorts to various alluring tricks to decoy the intruder away from her eggs.

REDSTART.

Description of Parent Birds. — Length about five and a quarter inches. Bill of medium length, straight, and black. Irides hazel; forehead white; crown, nape, back, scapulars, and wing-coverts deep bluish-grey; wing-quills brown; rump, upper tail-coverts, and tail-quills bright rusty red, with exception of the two middle feathers of the tail, which are brown like the wings. Over the base of the upper mandible, chin, cheeks, throat, and sides of the neck black. Breast bright rust-colour, belly paler, vent yellowish-white, under the tail red. Legs, toes, and claws, dark brown.

The female lacks the black and white on the head; has the upper-parts greyish-brown, tail duller, and under-parts fainter.

Situation and Locality. — In holes in trees, walls, thatches, rocks, and ruins; occasionally in such situations as inverted plant pots and disused pumps; sparingly distributed over England, Wales, and Scotland, but rarely met with in Ireland. Our illustration was procured in Norfolk.

Materials.—Dead grass, rootlets, and leaves, with an inner lining of hair and feathers.

Eggs.—Four to six, occasionally eight, of a pale bluish green, unspotted, and polished. Some text-books say, sometimes with a few faint red specks, but I have never found a clutch of eggs marked in any way. Size about ·75 by ·54 in. Situation of nest prevents confusion with Hedge Sparrow.

Time.—May, June, and July.

Remarks.—Migratory, arriving in April and leaving in September. Notes: song, pleasing and imitative, uttered very late in the evening and very early in the morning whilst the female is sitting; call, represented by some authorities as *chippoo*, and others as *oirhit*. Local and other names: Red-tail, Fire-tail. Bran-tail, Fire-flirt. Sits very closely indeed.

RING OUZEL.

Description of Parent Birds.—Length about eleven inches. Bill brownish-black, with a variable amount of yellow at the base, nearly straight, and of medium length. Irides hazel. Head, neck, back, wings, rump, and tail black, slightly tinged with brown, and margined, more or less, with grey, especially on the wings. Chin, throat, and under-parts blackish-brown, the feathers being bordered with grey. Across the breast is a broad, curved band or crescent of white, edged with a brownish tint. The legs, toes, and claws are brownish-black.

The female is browner and greyer, and the crescent on her breast much duller and less defined.

Situation and Locality.—In clefts of rock, steep banks, holes in stone walls, barns, limekilns, and sometimes quite on the ground, in the mountain and moorland parts of the North and West of England, Scotland. Wales, and Ireland. The illustration on page 233, procured in Westmoreland, is a very typical example of the situation of the nest.

Materials.—Small twigs, roots, coarse grass. moss, and mud, with an inner lining of fine grass. It is a very similar structure to that of the Blackbird, but generally found in more lonely and secluded districts.

REDSTART.

Eggs.—Four to five, blue-green, freckled, and spotted with brown. They are, as a rule, covered with larger spots than the eggs of the Blackbird, but upon occasion the latter will lay eggs resembling them so closely that it is quite impossible to distinguish without seeing the parent birds or knowing something of the locality of the nest. Size about 1·2 by ·84 in.

Time.—April, May, and June, generally the last two months.

Remarks.—Migratory, arriving in April and departing about the end of October. Notes: song, desultory, plaintive, and far-sounding. Local and other names: Rock Thrush, Ring Thrush, Rock Ouzel, Tor Ouzel, Ring Blackbird. Sits pretty close, and is somewhat demonstrative when disturbed.

ROBIN.   *Also* REDBREAST.

Description of Parent Birds. —Length about five and three quarter inches. Bill of medium length, nearly straight, and black; crown, nape, back, wings, and tail, olive-brown. Round the base of the beak, eyes, and upon the throat and upper breast, orange red, succeeding which is a narrow space of bluish-grey; the rest of the under-parts white, tinged with brown on the sides, flanks, and under tail-coverts. Legs, toes, and claws reddish-brown.

The female is slightly smaller, and her coloration is not quite so bright.

Situation and Locality.—In a hole in a bank. Our second illustration shows one in a typical situation, occupied by a young Cuckoo at the time it was photographed; and our first, one just under

RING OUZEL.

the roof in a corner of a cart-shed and tool-house which was used daily. The nest of the Robin has been found in every conceivable situation—holes in walls inside and outside buildings, in flower-pots, old boots, teapots, canisters, hats, and upon one occasion, at least, in a human skull. Common nearly everywhere throughout the British Isles.

Materials.—Fibrous roots and moss lined with dead leaves and hair.

Eggs.—Five to six, occasionally as many as seven and even eight; white or very light grey, blotched and freckled with dull light red. Sometimes the spots become confluent over nearly the entire surface of the shell, at others they are collected round the larger end. Occasionally very sparingly supplied or altogether absent. Size about ·8 by ·6 in.

Time.—March, April, May, June, and July.

Remarks. — Resident and migratory. Some naturalists are of opinion that the Robins which inhabit our gardens and orchards in winter migrate North in summer, and that their places are supplied by more Southern members of the species. Any-way, it is certain that the bird does migrate, from the fact that specimens visit our lightships during the great autumn rushes. Notes: call, sharp and clear; alarm, a very monotonous, low and plaintive *chee*, hardly ever uttered except when the nest is being visited by an intruder. Song, sweet and plaintive. Local and other names: Redbreast, Robin Redbreast, Robinet, Bob Robin, Ruddock. Sits closely, yet though a bold bird, will sometimes forsake its discovered eggs in the most unaccount-able manner. The situation represented in our first illustration has been occupied two or three years in succession.

ROBIN'S NEST IN THE CORNER OF A TOOL SHED.

ROBIN'S NEST IN A HEDGE BOTTOM, CONTAINING A
YOUNG CUCKOO.

ROOK.

Description of Parent Birds. — Length from eighteen to twenty-one inches. Beak large, strong, arched towards the point, and black. Round the base of the bill in the adult bird the skin is bare, scurvy, and light grey. This feature readily distinguishes it from the Carrion Crow. Irides dark brown. The whole of the plumage is black, glossed with rich purple on the upper-parts. Legs, toes, and claws black.

The female is, as a rule, smaller and less brilliant.

Situation and Locality.—Amongst the highest branches of tall trees, in colonies or rookeries of various sizes, throughout the country. I have seen colonies of a dozen birds in an isolated clump of ash trees away up in bleak hilly districts, and as a contrast to this, it may be mentioned that in 1847 it was computed that Newliston Rookery, near Edinburgh, contained no less than 2,663 nests.

Materials.—Sticks and twigs knitted and plastered together with mud and clay, and lined with straw, hay, or wool. The bird is often very particular about the kind of nest it constructs, and will pull it to pieces and rebuild it several times. It is an arrant rogue, and I have watched individuals steal each others' sticks. The old nests are sometimes repaired in the autumn, and it is said eggs are laid. Our illustration is from a photograph taken in Westmoreland, and shows a cluster of two or three nests built into each other.

Eggs.—Four to five, of a pale green or brownish-green ground colour, spotted and blotched with greenish or smoky-brown. Average size about 1·68 by 1·18 in. Distinguished by bird's gregarious habits.

ROOKS' NESTS.

Time.—February, March, April, and May; the laying season varying according to the character of the weather.

Remarks.—Resident. Notes, *craae*. Local and other names: White-faced Crow, Craa. Gregarious, and a close sitter.

RUFF.

Description of Parent Birds.—Length about twelve and a half inches. Bill long, straight, rather slender, and brown. Irides dusky brown. The bird varies very considerably in plumage, one eminent authority having examined two hundred specimens and only found two alike. Yarrell says: "The head, the whole of the ruff or tippet (long plumes growing on the head and neck, and capable of being raised so as to form a kind of shield), and the shoulders of a shining purple-black, transversely barred with chestnut; scapulars back, lesser wing-coverts, and some of the tertials, pale chestnut speckled and tipped with black; greater wing-coverts nearly uniform ash-brown; quill-feathers brownish-black, with white shafts; rump and upper tail-coverts white; tail feathers ash brown, varied with chestnut and black; the feathers of the breast, below the ruff and on the sides, chestnut, tipped with black; belly, vent, and under tail-coverts white, with an occasional spot of dark brown; legs and toes pale yellow-brown; claws black."

The female is about two inches less in length, lacks the ruff or tippet altogether, and although not differing much in other plumage, is said to be more uniform in colour as a sex.

Situation and Locality.—In a tuft or tussock of some kind of coarse vegetation growing in some wet, swampy place. The bird used to breed at several places in England, but the reclamation of land, the rquirements of gourmands, and, later, the greed of the collector, have almost banished it as a breeding species. It is now only known to attempt to breed in Norfolk and Lincolnshire, and it is doubtful whether the bird will long essay the almost hopeless task.

Materials.—A few bits of dead grass or leaves, line the hollow in which the eggs are laid.

Eggs.—Four, varying from pale greyish-green to olive-green or olive-brown in ground-colour, blotched and speckled with greyish and rich liver-coloured brown, generally most numerous on the larger end of the egg. Size about 1·7 by 1·22 in. Somewhat similar to those of the Redshank, though greyer and not quite so yellow in ground colour.

Time.—May and June.

Remarks.—Migratory, arriving in April and May and departing about September. Notes, *kack, kick.* Local and other names: Reeve (female), Fighting Ruff, Shore Sandpiper, Greenwich Sandpiper, Yellow-legged Sandpiper, Equestrian Sandpiper. Sits closely.

SANDPIPER, COMMON.

Description of Parent Birds.—Length seven and a half inches. Bill rather long, straight, slender, dark brown towards the tip and lightish brown at the base. Irides dusky brown; from the base of the beak a light streak runs over the eye and ear-coverts. Crown, back of the neck, back, wing-coverts, and upper tail-coverts greenish-brown, with a line of a darker hue across and down the centre of each feather. Wing-primaries nearly black,

with dirty white patches on nearly all the inner
webs, the secondaries tipped with white. Tail-quills
greenish-brown in the centre, barred with greenish-
black, the outer webs of the two outside feathers
on either side white, barred with greenish-black.
Chin, throat, breast, belly, vent, and under tail-
coverts white ; sides of the neck and upper portion
of the breast duller, and streaked with dark brown
or dull black. Legs and toes pale bluish-green ;
claws dark brown.

The female resembles the male.

Situation and Locality.—On the ground ; in a
hole in a bank, under the shelter of a tuft of grass,
in a tuft of rushes ; sometimes in a slight declivity
on the bare ground, or in a patch of grass amongst
large stones on a little river island : on the banks
of rivers, mountain streams with rough, gravelly,
and rocky banks, lakes, tarns, and reservoirs, in the
extreme South-west of England (Cornwall, Devon,
and Somerset), Wales, Derbyshire, the six northern
counties Scotland, and its surrounding islands,
and Ireland. Our illustration was procured in
Mull.

Materials.—Short pieces of dead rushes, sometimes dead leaves, with an inner lining of fine dry
grass.

Eggs.—Four, pale straw to creamy-yellow in
ground colour, with dark brown spots and blotches
on the surface, and underlying markings of light
brown and grey. Size about 1·5 by 1·08 in.

Time.—May and June.

Remarks.— Migratory, arriving in April, and
departing in September, although individuals may
be seen later. Notes : *wheet, wheet, wheet.* Local
and other names : Summer Snipe, Sand Lark, Willy
Wicket, Sand Lavrock, Spotted Sandpiper. Sitting

COMMON SANDPIPER.

Q

qualities variable, some individuals sitting closely and others lightly, irrespective of the condition of the eggs.

SANDPIPER, WOOD.

The appearance of this bird, even as a visitor, is neither frequent nor regular. It has been found breeding with us only twice, at the outside, during the last forty years, so that it does not call for special attention here.

SCOTER, COMMON. *Also* BLACK SCOTER.

Description of Parent Birds.—Length about twenty-one inches. Bill of medium length, swollen into a knob at the base, and flattened at the tip. It is black, with the exception of a ridge of yellow, which commences half an inch from the tip and extends to the base. Irides dusky brown. The plumage is deep black all over, somewhat glossy about the head and neck. Legs, toes, and webs dusky, darkest on the last.

The female lacks the knob on the bill, and her plumage is duller.

Situation and Locality.—A hollow scraped in the ground or some natural declivity, hidden by low, growing shrubs or sheltering heath ; on small islands near lochs and rivers, not far from the sea, in the most northern counties of Scotland.

Materials.—Twigs, heather, stalks, dead leaves, and dry grass, with an inner lining of down. The tufts are brownish-grey with pale centres, are large, a little darker than those of the Mallard, and much more so than those of the Goosander.

Eggs.—Six to nine, pale greyish-buff or yellowish-white, sometimes slightly tinged with green, smooth surfaced. Size about 2·5 by 1·78 in.

Time.—May and June.

Remarks.—A winter visitor principally. Is said not to breed until it is two years old. Notes: call, a grating *kr-kr-kr*. One authority represents the call-note of the male'as *tü-tü-tü-tü*, and the response of the female as *re-re-re-re-re*. Local and other names: Black Duck, Black Scoter, Black Diver. Sits closely, and covers its eggs when leaving them voluntarily.

SCOTER, BLACK. *See* SCOTER, COMMON.

SHAG.

Description of Parent Birds.—Length about twenty-seven inches; bill rather long, hooked at the tip, and black. The gape extends behind the line of the eye, and the naked skin about it is black, spotted with chrome-yellow. Irides green, The forehead bears a curved-forward kind of crest, which makes its appearance early in the spring. Head and neck, all round, rich dark green, glossed with purple and bronze sheen; back and wing-coverts dark green, with a more intense margin of the same colour round the border of each feather; wings and tail black; breast, belly, and under-parts generally the same as the head and neck; legs, toes, and claws black.

The female resembles the male.

Situation and Locality.—Crevices, fissures, and caves in sea cliffs; sometimes on ledges or amongst

the boulders of the rock-strewn beaches of small
islands. Pretty generally round our coasts, where
suitable accommodation is to be found, but principally
on the west coast of Scotland. The illustration
on p. 59 represents a cave in which a number of
pairs of this bird, Rock Doves, and a pair of
Herons were breeding in the outer Hebrides.

Materials.—Seaweed and twigs, lined with grass,
the whole plastered and befouled with droppings
and decomposing fish. Where conditions admit, it
is a bulky structure.

Eggs.—Two to five, generally three; pale green,
almost wholly covered by a chalky substance, which
soon becomes discoloured. The eggs resemble those
of the Cormorant closely, but are usually a trifle
smaller in size. The situation of the nest and
presence of birds readily distinguish them. Average
measurement 2·45 by 1·5 in.

Time.—May and June.

Remarks.—Resident, but subject to local movement. Note, a harsh guttural croak. Local and other
names : Crested Shag, Crested Cormorant, Green
Cormorant. Shag and Cormorant are names frequently interchanged by seamen and coast dwellers.
Gregarious. A bold and fairly close sitter.

SHEARWATER, MANX.

Description of Parent Birds.—Length about
fourteen inches ; bill rather long, straight, except at
the tip, where it is curved downwards, and blackish-
brown, lighter at the base. Irides hazel. Head,
nape, back, wings, and tail brownish-black ; chin,
throat, breast, belly, vent, and under tail-coverts
white. The sides of the neck are barred transversely

with grey and white. A patch of brownish-black is situated behind the thigh on each side ; legs, toes, and webs flesh-colour, tinged with yellow.

The female is similar to the male, but slightly smaller.

Situation and Locality.—At the end of a burrow, generally excavated by the bird, and varying from two to ten or twelve feet in depth, in crevices, and under pieces of rock ; sometimes in a small hole scratched out between two stones. In the Scilly Islands, on the islands to the west of Scotland, and in suitable places off the Irish coast. It is possible that its peculiar habit of keeping out of sight during the day and only coming forth at night may have conduced to some of its nesting haunts having been overlooked. The bird is known in one case to have been driven away from its nesting stations by Puffins, and in another by rats.

Materials.—Sometimes a few dead fern-fronds, or blades of dried grass, at others nothing whatever.

Egg.—One ; pure white, smooth, and large for the size of the layer. Size about 2·4 by 1·65 in.

Time.—May and June.

Remarks.—Resident, but subject to much local movement, except during the breeding season. Notes, said to be a guttural melody, delivered in their holes, and resembling *crew cockathodon.* Local and other names : Shearwater Petrel, Manx Puffin, Cuckle, Skidden, Scraib, Fachach, Lyrie, Scrapire. Gregarious. A close sitter.

SHELDRAKE, COMMON. *See* SHELDRAKE.

SHELDRAKE. *Also* COMMON SHELDRAKE *or* BURROW DUCK.

Description of Parent Birds.—Length twenty-four to twenty-six inches; bill fairly long, thick at the base, depressed in the middle, slightly hooked at the tip, and red in colour. On the top of the upper mandible, at the base, is a large fleshy knob. Irides reddish-brown. Head and upper parts of neck dark green; lower half of neck white all round; upper parts of breast and back, rump, wing, and upper tail-coverts white; scapulars and a portion of secondaries blackish; outer webs of inner secondaries rich chestnut. On the last-named feathers is a patch of rich purple green; primaries almost black; tail-quills white, except at the tips where they are black; in the middle of the lower breast and belly the feathers are dark brown; sides, flanks, vent, and lower tail-coverts white; legs, toes, and webs flesh-colour.

The female is rather smaller and duller in colour.

Situation and Locality.—Rabbit burrows are the favourite haunts of this bird, although it is said sometimes to dig its own burrow or adopt that of a fox or badger; holes under rocks and ruins at various depths, sometimes four or five feet in, at others as many as twelve. In low sand-hills and dunes at various suitable places on the east and west coasts, such as Suffolk, Norfolk, Lincolnshire, Yorkshire, Durham, Northumberland, Lancashire, and Cheshire; on various parts of the coast of Scotland, Orkney Islands, Hebrides, and Ireland. Our illustration was obtained in the Hebrides.

Materials.—Dry grass, bents, and down from

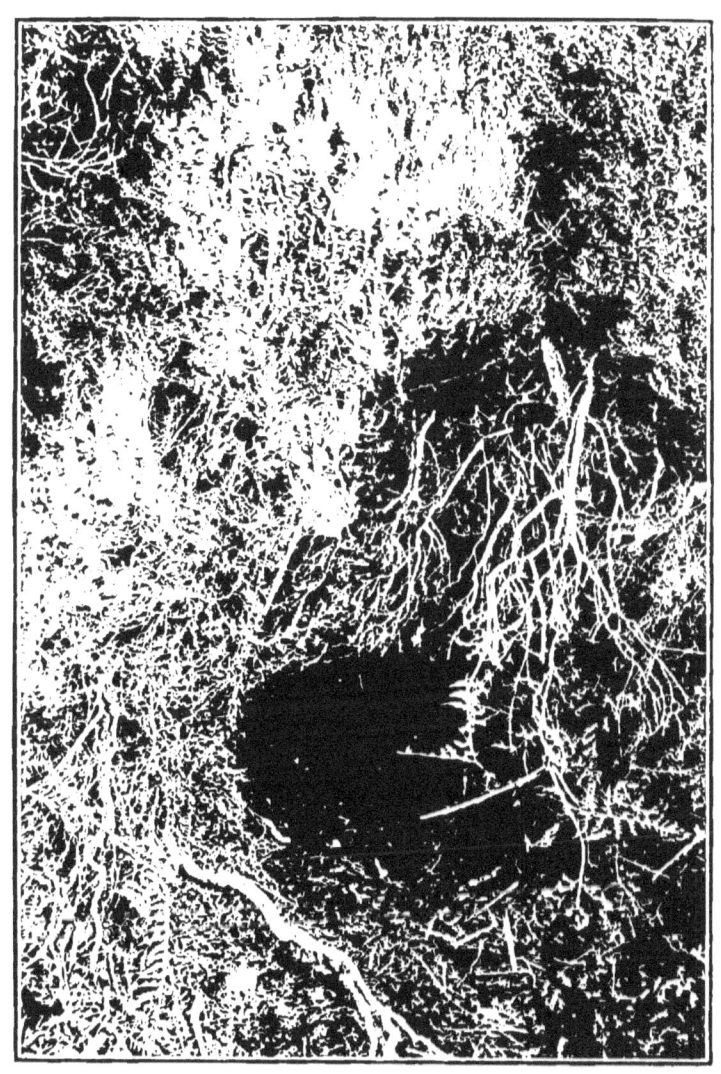

SHELDRAKE.

the bird's own body. The tufts are lavender-grey, mixed with a few white ones.

Eggs.—Six to sixteen, generally ten to twelve; white, slightly tinged with cream colour. Size about 2·7 by 1·9 in. Nest distinguished by down-tufts.

Time.—April, May, and June.

Remarks.—Resident. Notes : male call, a deep *korr-korr ;* female, a loud quack. Local and other names : Sly Goose, Bargander, Burgander, Burrow Duck, Skeeling Goose, Common Shieldrake, Skelgoose. The nest is very difficult to find ; but a good way to accomplish this is to look out for footprints in the sand at the entrance to likely holes, also to watch the movements of the male during flight, and any suspected hole morning and evening when the birds leave and enter. I have generally been astir at three o'clock in the morning for that purpose, and the subject of our picture was discovered not long after that hour by the aid of my binoculars. Sits closely.

SHOVELLER.

Description of Parent Birds.—Length about twenty inches. Bill rather long, narrow in the middle, and widening towards the tip, nearly straight, and leaden grey. Irides yellow. Head and upper part of neck deep glossy green, lower part of neck white. Back, in the centre, blackish-brown, the feathers being edged with a lighter tint. Lesser wing-coverts light blue, greater white ; scapulars white, quills brownish-black. The speculum, or glossy patch upon the wing, is green. Rump, upper tail-coverts, and tail-feathers brownish-black. Breast and belly chestnut ; vent white ; under tail-

SHOVELLER.

coverts black. Legs, toes, and webs reddish-orange ; claws black.

The female differs considerably from the male, having her head and neck mottled with two shades of brown ; the feathers on the upper surface of the body are dark brown in the centre, bordered with a lighter shade of the same colour. Under-parts of the body pale brown.

Situation and Locality.—In tufts of grass, rushes, and heath beside sluggish rivers, lakes, broads, tarns, and swampy heaths, in favourable situations on the eastern coast of England ; also in Scotland and Ireland. Our illustration is from a photograph taken in Norfolk.

Materials.—The little hollow chosen is lined with sedges and dead leaves, dry grass, and, as incubation advances, down. The tufts are of a dark neutral grey colour, lighter in the centre and tipped with white. They and the female form reliable distinctions.

Eggs.—Seven to ten, or occasionally fourteen, buffish-white, tinged with green, unspotted, and slightly polished. Closely resemble those of the Mallard and Pintail. Size about 2·0 by 1·5 in.

Time.—May.

Remarks.—Resident and migratory, being more numerous in winter than in summer. Notes : *quack*, uttered in deeper tones by the male; when flying, *puck*, *puck*. Local and other names : Broad Bill, Blue-winged Shoveller. Sits closely.

— —

SHRIKE, RED-BACKED.

Description of Parent Birds.—Length about seven and a half inches ; bill rather short, hooked

RED-BACKED SHRIKE.

at the tip, and black or dusky brown, lighter at
the base of the lower mandible. Near the end of
the upper mandible is a prominent tooth or notch.
Irides hazel. Three or four strong black bristles
spring from just above the gape. Round the base
of the upper mandible, through the eyes, and as far
as the ear-coverts, the feathers are black ; crown and
back of neck grey ; back and wing-coverts bright
reddish-brown ; wing-quills dull black margined with
reddish - brown; upper tail - coverts reddish - grey;
tail-quills, in centre, black tipped with white, rest
white on the basal half, and black from thence to
the end, which is slightly tipped with white ; chin
grey; breast, belly, and under-parts of a rosy
tinge, with the exception of the under tail-coverts,
which are white ; legs, toes, and claws dusky
black.

The female is very much less conspicuous in
her plumage. Her beak is not so dark in colour;
over her eye is a yellowish-white streak ; her
upper-parts dull rusty brown, tinged with grey on
the neck and tail-coverts ; chin, throat, breast,
and under-parts greyish-white, barred with greyish-
brown.

Situation and Locality.—In high rough hedges,
thorn bushes in woods, and on rough commons.
Our illustration is from a photograph of a nest in
a slight thorn bush, surrounded by hazels and big
trees in a small Surrey spinney, where I meet with
a nest every year regularly. I found a Red-backed
Shrike's nest in a low bramble bush, intermixed
with rushes, two years ago close to London. The
nest was not more than eighteen inches from the
ground, and within a few feet of a much-used turn-
pike lane. The cock bird was so bold that he
came within four feet of me as I stood looking at

his mate sitting upon her eggs. Breeds pretty generally over England, with the exception of the extreme north, but is rarely met with in Scotland or Ireland.

Materials.—Slender twigs, dead grass, stalks, dead weeds, honeysuckle stems and stalks, roots, wool, moss, and sometimes feathers, lined with hair, sometimes with willow catkins and fine, fibrous roots. As a rule it is a very large nest for the size of the bird; but I have noticed that specimens differ in this respect as well as in the character of the materials employed in their construction.

Eggs.—Four to six, generally four or five; very variable in ground colour and markings; pale buffish-white, spotted, freckled, and blotched with pale reddish-brown, and underlying markings of grey or salmon colour, marked with light red and lilac-grey. Some varieties are white, greyish-white, yellowish-white, or greenish in ground colour. As a rule, the markings form a ring round the larger end. Size about ·9 by ·66 in.

Time.—May and June. I once found one at the beginning of July.

Remarks.—Migratory, arriving in May and departing in August or September. Note: call, a harsh croak; song, a mixture of the notes of the Goldfinch, Blackcap, Nightingale, and other birds frequenting its vicinity, according to Bechstein. Local and other names: Jack Baker, Murdering Pie, Whiskey John, Butcher Bird, Flusher, Cheeter. A close sitter.

SHRIKE, WOODCHAT.

A rare and accidental visitor, which is said to have bred once or twice in the Isle of Wight.

SISKIN.

Description of Parent Birds.—Length nearly five
inches. Bill short, conical, sharp-tipped, and orange-
brown. Irides dusky brown. Top of head black;
over and under each eye is a yellowish streak. Sides
of head yellowish-green; nape and wings (except
greater coverts and quills, which are brownish-
black, tipped and bordered with yellow) greenish-
olive, streaked with black; rump yellow; upper tail-
coverts greenish-olive. Tail slightly forked, and
dusky black, yellowish on the upper half, with the
exception of the middle pair of feathers. Chin
black, throat and breast yellowish-green; belly, sides,
flanks, vent, and under tail-coverts, greyish-white,
streaked with dusky black. Legs, toes, and claws
brown.

The female is smaller, and lacks the black on
the crown and chin. Her upper-parts are olive-
brown, throat and breast greenish-yellow, and rest
of under-parts greyish-white. With the exception
of the centre of the belly she is streaked all over
with dusky black.

Situation and Locality.—In plantations, woods,
and forests. Its nest has been found on very
rare occasions in different parts of England, in
furze and juniper bushes; but in Scotland, where
it breeds sparingly, it adopts higher situations
amongst the forks and branches of fir-trees.

Materials.—Slender twigs, dried grass, and moss,
lined internally with hair, rabbit or vegetable down,
and sometimes a few feathers.

Eggs.—Four to six, greyish-white, tinged with
green or pale bluish-green, spotted and speckled
with rusty and dark brown spots, sometimes streaked

with the darker colour. The markings are generally scattered over the surface of the eggs, but are sometimes collected round the larger ends. They resemble those of the Goldfinch very closely indeed, but are said to run larger, and the ground colour to be of a darker tinge. The situation of the nest and a sight of the owner are the only reliable evidences, however. Size about ·66 by ·52 in.

Time.—April, May, and June.

Remarks.—A winter visitor of erratic appearance; a few resident. Notes: call, a metallic *keet*; alarm note, *chuck-a-chuck, keet.* Some naturalists represent the call-note as a weak *tit-tit-tit-tit,* and *tsyzing,* others as a loud *deedel* or *deedlee.* Local and other names: Aberdevine (used by bird-catchers), Barley Bird. Nest difficult to find. A close sitter.

SKUA, COMMON. *Also* SKUA, GREAT.

Description of Parent Birds.—Length about twenty-four inches. Bill of medium length, hooked at the tip, and with bare skin round its base, black. Irides dark brown. Head and neck dark umber brown, slightly streaked with lighter brown; back, wings, and tail-coverts dark brown, streaked with light reddish-brown. In some specimens the feathers at the nape, and the middle and edges of those on the back, are greyish-white. The wing-quills are white at the base and blackish-brown towards the tip; tail-quills very dark brown. Chin and front of neck, breast, belly, vent, and under tail-coverts dusky rust colour. Legs, toes, and webs black; claws large, much curved, strong, and black.

Situation and Locality.—On the ground, amongst moss or heather; in the Shetland Islands only, where, in spite of protection, we were told, during our visit to the outer Hebrides, that not a single young bird was reared during 1894.

Materials.—Sometimes a few bits of grass are used as a lining; at others nothing is placed in the hollow or cavity. Some nests are said to be made of dead ling, moss, and dry grass.

Eggs.—Two, occasionally only one, varying from light buff to dark olive-brown, blotched and spotted with dark brown and rusty or greyish-brown. Size about 2·85 by 1·95 in. Similar to Lesser Black-backed and Herring Gulls, but markings are fewer and duller, and presence of parent birds readily settles the point.

Time.—May and June.

Remarks.—Migratory, arriving at its breeding haunts in April and leaving in August. Notes: *ag-ag* and *skua*. Local and other names : Great Squa, Bonxie, Brown Gull, Skua Gull, Morrel Hen. Gregarious. It is said by one observer to prepare several nests before deciding in which to drop its eggs. Sits lightly.

SKUA, GREAT. *See* SKUA, COMMON.

SKUA, RICHARDSON'S.

Description of Parent Birds.—Length about twenty inches. Bill moderately long, strong, straight, except at the tip, where it is hooked, bluish lead-colour at the base, and blackish elsewhere. Irides dark brown. This bird is subject to considerable individual variation, and there are two distinct and well-marked varieties, known as "light" and "dark," which interbreed freely. The dark variety is more common in low latitudes, and the light one in high latitudes, as might be expected.

Mr. Seebohm, in describing the bird, says:—' In the adult of the dark form, the whole of the plumage is an almost uniform dark sooty brown, slightly suffused with slate-grey on the upper-parts, and with a bronzy yellow on the sides of the neck.

" In the adult of the light form, the slate-grey of the upper-parts is a little more pronounced than in the dark form. The general colour of the under-parts is white, shaded with brown on the sides of the breast, the vent, and the under tail-coverts; the white on the throat extends round the sides of the neck and across the lower ear-coverts, almost to the nape, and is suffused with yellow. Legs and feet black."

The female is, so far as is known, indistinguishable from the male, except that the elongated feathers of the tail are somewhat shorter.

Situation and Locality.—On the ground, amongst heather, moss, and coarse grass, in the moorland parts of the Orkneys, Shetlands, Hebrides, and in one or two places on the extreme northern mainland of Scotland. Our illustration is from a photo-graph taken in the outer Hebrides.

Materials.—Dried grass and moss, used as a scanty lining to the hollow in which the eggs are laid; sometimes none used at all.

Eggs.—Two; as many as three have been found, and, upon occasions only one. Varying in ground colour from dark olive-green to brownish-green, irregularly spotted and blotched with differing shades of dark brown and greyish-brown, generally dis-tributed over the entire surface of the egg, but sometimes most numerous at the larger end. They

R

closely resemble some of the Gulls' in appearance,
and the only safe method of identification is to
watch the parent bird on or off the nest. Size about
2·3 by 1·62 in.

Time —May and June.

Remarks.—Migratory, arriving in May and departing in August and September. Notes: *mee* and
mee-awk, represented by some authorities as *kyow*
and *yah-yah*. Local and other names: Arctic Gull,
Arctic Skua (also applied to Long-tailed or Buffon's
Skua), Shooi, Scoutie-Allen, Black-toed Gull. Sits
lightly, and is gregarious.

SKYLARK.

Description of Parent Birds.—Length about seven
inches. Bill of medium length, straight, strong, and
dark brown. Irides hazel; crown dark brown, the
feathers being edged with a lighter and redder tinge,
and somewhat elongated, forming a crest which is
erectable at will. Back of neck, back, wing and tail-
coverts reddish-brown, each feather being bordered
with a pale tint. Wing and tail-quills dusky brown,
with lighter edges and tips. Throat and breast light
cream-colour, spotted with dark brown; under-parts
pale straw-colour, tinged with brown on the thighs
and flanks. Legs, toes, and claws brown; middle
toe largest, and hind claw very long and curved.

The female is not quite as large as the male, but
is similar in her plumage.

Situation and Locality.—Under tufts of grass,
ling, and heath, sometimes on the plain open
ground, in a slight declivity. Our illustrations
are from photographs of a nest on the crown of a
furrow, and under a tuft of grass. The former was

RICHARDSON'S SKUA.

procured in Surrey and the latter in Yorkshire. In cultivated and uncultivated districts throughout the United Kingdom, but not in woods and plantations.

Materials.—Grass, roots, and horsehair, the latter two often quite absent and the first used sparingly.

Eggs.—Four to five, of a dirty white ground colour, occasionally tinged with olive-green, thickly spotted and speckled with olive-brown, and under-lying markings of greyish-brown. The markings are generally so thickly and evenly distributed as to hide the ground colour, but occasionally, the mark-ings are less thickly distributed and collected in a kind of belt at the larger end of the egg. Size about ·93 by ·68 in. Distinguished from Woodlark by crowded olive-brown markings.

Time.—April, May, June, and July.

Remarks.—Resident, though subject to partial migration and much local movement. Notes : song consists of several strains, trilling, warbling notes, variously modulated, and interrupted now and again by loud whistling. Local and other names : Lavrock, Field Lark. A close sitter when the ground is rough and uneven, but not particularly so when it is bare and the situation exposed.

SNIPE, COMMON.

Description of Parent Birds.—Length about ten and a half inches ; beak very long (about two and three-quarter inches), straight, and pale reddish-brown at the base and dusky towards the tip. Irides dark brown. Crown blackish-brown, divided in the centre by a buffish-brown longitudinal line ; another line of the same colour commences at the base of

SKYLARK'S NEST ON THE CROWN OF A FURROW.

SKYLARK'S NEST IN A ROUGH GRASS PASTURE.

the bill and passes over the eyes. A dusky line passes from the base of the beak to the eye; back and scapulars dark brown, barred with rusty brown. Four distinct lines of dark brown feathers, bordered with rich buff, run along the upper parts of the body. Wing-coverts dull black, spotted with pale brown and tipped with white; quills dull black, some of them edged and others tipped with white; upper tail-coverts dusky black, barred with brown; tail-quills black, barred and spotted with dull orange-red, and tipped with pale reddish-yellow; chin brownish-grey; neck and cheeks light brown; front and sides of neck a mixture of dark and rusty brown; breast, belly, and vent white; under tail-coverts pale brown, barred with dusky black; legs and toes greenish-brown, dusky, or leaden colour.

The female is practically like the male, except that she is a trifle larger.

Situation and Locality.—On the ground, in a tuft of long coarse grass, amongst rushes or heather, generally hidden by an overhanging tuft of half-dead grass. In wet pasture-lands, marshes, and swamps, near tarns and bogs, in suitable localities throughout the United Kingdom. Our illustration is from a photograph taken in Norfolk.

Materials. — A few dry grass stalks, slender sprigs of dead heather or other bits of herbage, used as a lining; sometimes hardly anything at all.

Eggs.—Four; ground colour varying from olive-green to greyish-yellow; spotted and blotched with blackish-brown, light brown, and underlying markings of grey. The markings are generally most numerous at the larger end, and the eggs are sharply pointed at the smaller. Size about 1·58 by 1·1 in.

Time.—April and May, although nests with

COMMON SNIPE.

eggs in have been found as early as March 20 and
as late as July 28.

Remarks.—Resident, but subject to local migration. Notes: *tjick-tjuck, tjick-tjuck,* uttered both
whilst the bird is perched and between the drum-
mings when it is on the wing at dusk and on dull
days. Local and other names: Hammer Blate,
Whole Snipe, Heather Bleater. Sits closely, and
simulates lameness when flushed, in order to draw
the intruder away from its eggs.

SPARROW, COMMON. *Also* SPARROW, HOUSE.

Description of Parent Birds.—Length nearly six
inches. Bill short, thick at the base, strong, and
dusky. Irides hazel. Crown and nape ash-grey;
back, scapulars, and wing-coverts reddish-brown,
mixed with black, the last-named feathers being
tipped with white, which forms a bar across the
wing. Wing-quills dusky, edged with reddish-brown;
tail dusky brown, bordered with grey. Cheeks
whitish; chin and throat black; belly and vent
light ash-grey. Legs and toes brown; claws black.

The female is not quite so large as the male;
the plumage on her upper-parts is not so bright,
and she lacks entirely the black on the chin and
throat. Town-dwelling birds have their plumage
much dulled by grime and smoke, which even
penetrates to the interior of their bones.

Situation and Locality.—Holes in walls, under
tiles and slates, behind signboards fastened against
walls, holes in cliffs, the old nests of Sand and
House Martins; in the thatch of houses, barns, and
ricks, holes in hollow trees; amongst ivy trained
against houses; the rafters of stables and sheds;

DOVECOTE IN WHICH SPARROWS NESTED.

amongst the loose sticks under Rooks' nests, in old
Magpies' nests, Pigeon cotes, on ledges and almost
any situation capable of accommodating a few straws
and feathers. I know several small villas in the
North of London with globular cast-iron ornaments
on their summits; into these the Sparrows have
found their way and turned them to account as
nesting sites. I once met with a colony in a low
whitethorn hedge, quite away from any houses
whatever, and have counted as many as twenty-six
nests in the branches of a single tree. Our illustration represents one of two Pigeon cotes, standing close together, out of which over two hundred
Sparrows' eggs were taken during the spring and
summer of 1894.

Materials. — Straw, hay, bits of string, moss,
worsted, and cotton rags, wool and hair, with a
liberal inner lining of feathers. Where the nest is
under cover it is not so bulky, and is open at the
top, as a rule; but where it is exposed it is covered,
bulkier, and better constructed, with a hole in the
side, and generally near the top.

Eggs.—Four to seven, generally five or six, pale
grey or greyish white, sometimes tinged slightly with
green or blue, spotted and blotched thickly with
brown, of various shades, and grey. I have some
Sparrows' eggs pure white. One egg generally
differs from the others in a clutch in regard to
the character of its markings. Size about ·9 by
·6 in. Distinguished from Tree Sparrow's eggs by
larger size, situation, and female lacking black
patch on her chin.

Time.—March to July or August.

Remarks.—Resident. Notes: a monotonous
chirrup, and a hurried scolding when engaged in
warfare. One authority makes the surprising state-

ment that when under proper training the bird can be taught to sing even better than the Canary. Local and other names: none. Sits pretty closely.

SPARROW, HEDGE. *See* HEDGE SPARROW.

SPARROW, HOUSE. *See* SPARROW, COMMON.

SPARROW, REED. *See* BUNTING, REED.

—

SPARROW, TREE.

Description of Parent Birds.—Length about five and a half inches. Bill short, strong, broad at the base, and lead coloured. Irides hazel. Crown and nape dull chestnut-brown. Beneath the eye is a streak of black. Cheeks white, with a large black spot. The upper part of the back and scapulars bright rusty brown, streaked with black; lower part of back and upper tail-coverts brownish-grey. Lesser wing-coverts bright rusty brown; greater black, with rusty-coloured edges and white tips; quills dull black, bordered with rusty brown. Tail-quills greyish-brown, edged with lemonish-brown. Chin and throat black; sides of neck, running somewhat far back, white; breast bright ash-grey; belly dull white, tinged with buffish-brown on the sides, vent, and under tail-coverts. Legs, toes, and claws pale yellowish-brown.

The female is a little smaller, but her plumage differs in nothing but its lesser brilliancy from that of the male. In this respect the bird differs radically from the House Sparrow, the female of which is not adorned by the black patch on the chin and throat, so conspicuous in the male.

Situation and Locality.—Holes in pollards and other trees, crevices of rocks, holes in walls, and the thatch of barns. It is very local, and nowhere numerous. It breeds most commonly in the midland and eastern counties of England, and is met with sparingly in Scotland, Wales, and Ireland.

Materials.—Straws, dry grass, and roots, lined with hairs and feathers. It is generally open at at the top, but domed when situation demands it.

Eggs.—Four to six, generally four or five, greyish-white in ground colour, thickly spotted all over with dark grey or dark brown. Occasionally the ground colour is white, thickly spotted and freckled with grey spots and blotches. One egg of a clutch is often lighter coloured than the rest, and sometimes the eggs are streaked with a dark line or two. Size about ·8 by ·57 in. They are not unlike the eggs of the Pied Wagtail and Meadow Pipit, but of course the position of nest differs widely. (*See* HOUSE SPARROW, " Eggs.")

Time.—April, May, June, and July.

Remarks.—Resident, but subject to local movement. Notes : numerous Sparrow-like chirrups. Local and other names : Mountain Sparrow. A close sitter.

SPARROW-HAWK.

Description of Parent Birds. — Length about twelve inches ; beak short, curved, and bluish ; bare skin round the base of the beak yellow. Irides yellow. Head, nape, back, wing, and upper tail-coverts deep bluish-grey, edged with rusty red ; wing-quills dusky, barred with black on the outside

SPARROW-HAWK.

webs, and spotted with white on the lower portions
of the inside webs ; tail deep ash-colour, crossed
with broad bars of dull black and tipped with
whitish-grey ; throat, breast, sides, belly, and vent
reddish-brown, marked with transverse bars of orange
in some and brown in others; legs and toes yellow ;
claws black.

The female is about three inches longer, and
nearly twice as heavy. Her upper-parts are browner,
with the exception of the back of the head, which
is greyer. The breast and under-parts are lighter,
and the markings on them larger and browner,
Both sexes are subject to considerable variation,
and are said to grow greyer with age.

Situation and Locality.—In fir, alder, larch, oak,
pine, and other trees, in well-wooded districts
throughout the British Isles. It is generally
placed in a fork or on a strong horizontal
branch. Our illustration was procured in Westmoreland.

Materials.—Sticks and twigs, the finest in the
centre, which is simply a slight hollow on a large
platform. Many naturalists assert that the bird
often utilises the old nest of a Magpie or Crow ;
but Mr. Dixon says that the nest is always
made by the birds themselves. I have taken some
eight or ten nests personally, and in every single
instance I am able to endorse him.

Eggs. — Four to six, generally five ; ground
colour white, tinged with blue or bluish-green,
clouded, blotched, and spotted with pale brown
and dark rich brown. The markings generally
form a zone round the larger end of the egg ;
sometimes the ground colour is almost entirely
hidden, and at others nearly, if not quite, all
exposed. I have noticed in clutches of six eggs

two of them are invariably less marked than the remaining four. Size about 1·65 by 1·3 in.

Time.—April, May and June, depending somewhat upon a southern or northern district.

Remarks. — Resident. Note: alarm, a harsh scream. Local and other names: Pigeon Hawk. A close sitter. I have known the hen return to the nest and continue to sit even when badly wounded.

STARLING.

Description of Parent Birds.—Length about eight and a half inches. Bill rather long, nearly straight, and yellow, except at the base, where it is light bluish-grey. Irides brown. The head, neck, and upper-parts are black, glossed with purple-green and steel-blue; the feathers of the head and neck are very slightly tipped with buffish-white; those of the back, rump, and upper tail-coverts are tipped with larger spots of the same colour. Wing and tail-quills greyish-black, edged outwardly with buffish-white; breast and belly black, glossed with purple and steel-blue; vent and under tail-coverts black, tipped and edged with buffish-white, lighter than on the back. Legs, toes, and claws reddish-brown.

The female is not so bright as the male, either in her plumage or bill.

Situation and Locality.—Fissures and crevices in cliffs, holes in the gables of old houses, stables, and barns; in ruins, under eaves, in hollow trees, and sometimes even amongst the loose sticks forming the foundation of a Rook's nest. Our illustrations represent a hollow apple-tree, in which a pair of Starlings breed every year. During 1894, three clutches of six, six, and five eggs respectively, were

taken out of this nest, on account of the owner
of the orchard having seen the parent birds feeding
from the fruit of a strawberry-bed close by.

Materials.—Straw, hay, and fibrous roots, lined
with feathers, wool, moss, or whatever may be
easily obtainable. I have often seen nests made
with nothing whatever but straw.

Eggs.—Four to six, of a uniform pale blue. I
have seen clutches once or twice that were as near
white as possible. Size about 1·18 by ·84 in.

Time.—April, May, and June, although eggs
have been seen in January and later than June.

Remarks.—Resident, but subject to southern
movement in winter. Notes : alarm, *spate, spate;*
song, a mixture of all kinds of sounds, the bird being
a very clever imitator. Local and other names :
Sheep Starling, Stare, Sheep Stare, Solitary Thrush,
Brown Starling. A close sitter.

STONECHAT.

Description of Parent Birds.—Length a little over
five inches. Bill of medium length, slightly curved
downward, and black. Irides dusky brown. Head,
nape, and back black, edged with tawny brown ;
rump and upper tail-coverts white, tipped with tawny
brown and black. Wing-coverts black, edged and
tipped with rusty brown ; those nearest the body
are white, and form a conspicuous patch on the
wing ; quills dusky, some of them edged with
rusty brown. Tail-feathers black, faintly edged
and tipped with pale reddish-brown. Chin and
throat black ; sides of neck white ; breast dark rich
rust colour, belly much lighter ; vent and under
tail-coverts a mixture of black and white, which

STARLING.

S

varies in individual specimens; some are dark and others quite light coloured in these parts.

The female is dull brown on the head, nape, and back, the feathers being edged with buff; the rump is brownish, the chin buff, the sides of the neck brownish-white, and the breast and belly duller.

Situation and Locality.—On or near the ground, amongst grass, brambles, at the foot of gorse bushes, and amongst rough, tangled vegetation ; in pastures, grass fields, on furze and heath-covered commons, and ground covered with juniper brambles, boulders, and bushes. The nest is extremely difficult to find ; the example in our illustration was stumbled upon quite by accident on a Suffolk common. It is very local, but breeds more or less in suitable localities all over the British Isles.

Materials.—Roots, moss, and dry grass, with an inner lining of hair, feathers, finer grass, and sometimes a little wool.

Eggs.—Four to six, rarely seven, of a pale bluish-green, closely mottled, and especially round the larger end, with reddish-brown spots. Sometimes without any spots at all. Size about ·7 by ·57 in. Distinguished from the eggs of the Whinchat by lighter ground colour and more defined markings, also by parent birds.

Time.—April and May.

Remarks.—Resident, but subject to local migration. Notes : *ü-tic, ü-tic,* changing when the young are hatched to *chuck, chuck.* Local and other names : Stoneclink, Stone Chatter, Stone Smick, Stone Chack, Stonesmith, Chick Stone, Black Cap, Moor Titling, a name generally applied to the Meadow Pipit in some districts. A fairly close sitter, but when at the foot of a furze bush the bird runs for some distance before taking flight. It is

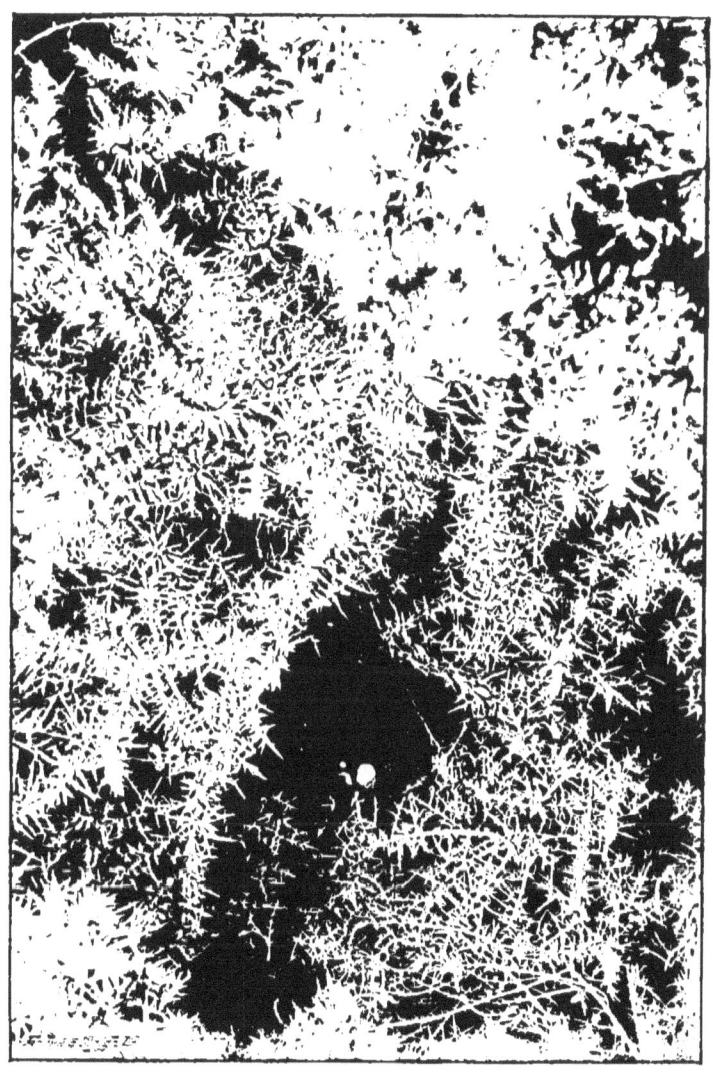

STONECHAT

extremely wary, and I have lain for hours and hours
together watching a pair through my binoculars
without being able to discover the nest.

SWALLOW.

Description of Parent Birds. — Length about
eight and a half inches; bill short, straight, somewhat flat, and black. Irides hazel. Forehead
chestnut; crown, neck, back, rump, and upper
tail-coverts steely blue; wings and tail sooty black,
the former long and sword-like, and the latter very
much forked From several specimens examined
I have found that one side of the tail (sometimes
the right and at others the left) is a trifle longer
than the other; chin and throat chestnut. Round
the lower part of the throat and upper part of
the breast is a broad steely-blue collar; lower breast,
belly, vent, and under tail-coverts buffy white;
legs, toes, and claws short, slender, and black.

The female is not so richly marked in her
plumage, and her tail is not so long.

Situation and Locality.—Generally, as shown in
our illustration, on the rafter of a barn, stable, or
shed. Sometimes on ledges and other projections
in chimneys and from walls. I recollect once
finding one inside an old disused mountain limekiln. We discovered several nests in a Surrey
bothy last summer that were built against the
whitewashed wall, and were exactly like those of
the Martin, except that the tops were open. Unfortunately, our photograph turned out a failure,
and when we returned to the district a few weeks
after, for the express purpose of securing a picture,
some farm boys had destroyed the nests.

SWALLOW.

Materials.—Mud, straws, dry grass, and feathers in liberal quantities.

Eggs.—Four to six, generally five, white, speckled and blotched with dark red-brown, and underlying markings of ash-grey. The markings are generally most numerous round the larger end. I have seen eggs once or twice with hardly any marking on at all. Size about ·83 by ·55 in.

Time.—May, June, July; and sometimes eggs may be met with as late as the beginning of August.

Remarks. — Migratory, arriving in April and departing in September and October. Notes, *wet-wet*, a warbling kind of song note, and *pink, pink* when the bird is alarmed. Local and other names: Barn Swallow, House Swallow, Chimney Swallow, Common Swallow. Not a very close sitter until incubation has advanced some stages.

SWAN, MUTE.

Description of Parent Birds. — Length, from about four feet eight inches to five feet; beak fairly long, black on the edges and tip, rest red. The knob, or tubercle, on the base of the upper mandible, and the naked skin between the eyes and the base of the bill, black. Irides brown. The whole of the plumage is snowy white. Legs, toes, and webs black.

The female has the knob smaller, the neck more slender, and swims deeper in the water.

Situation and Locality.—On the ground amongst reeds and coarse vegetation. On small islands and banks. Close to the water of sluggish rivers or lakes in various parts of the country, but principally on the Thames, Avon, Norfolk Broads, and at

MALE MUTE SWAN SITTING ON THE NEST.

Abbotsbury in Dorsetshire. Our illustration, representing the male bird sitting on the nest, was procured at Long Ditton.

Materials.—Reeds, rushes, dry flags, and grass, often in great quantities, and down.

Eggs.—Three to twelve, generally six or seven; dull greenish-white. Size about 4·5 by 2·9 in.

Time.—March, April, and May.

Remarks.—Strictly speaking, this bird has no proper claim for inclusion in a work of this character, for although it breeds in a perfectly wild state on the Continent, it has never been known to do so within the limits of the British Isles. The case of the Pheasant, however, another introduced half-domesticated bird, holding its own only through strict protection, paves the way.

The Mute Swan is said to have been first introduced into this country from Cyprus by Richard I., who commenced to reign in 1189. It is considered a bird royal when at large and unmarked, and is consequently afforded protection. Notes, soft and low, plaintive, and of little variety. Local and other names: Common Swan. A close sitter.

SWIFT.

Description of Parent Birds. — Length about eight inches; beak very short, with an extraordinary width of gape, and black. The whole of the plumage is a dingy black, except the chin, which is of a dirty white colour. The tail is of medium length and forked, and the wings very long and sword-like; legs, toes, and claws black. The feet have four toes, all of which are in front.

The female is similar to the male in size and

SURREY COTTAGE UNDER THE TILES OF WHICH A NUMEROUS COLONY OF
SWIFTS BREED YEARLY.

colour. Easily distinguished from Swallow by larger size, shorter tail, and colour.

Situation and Locality.—Holes in church towers, chimneys, sea cliffs, under the tiles of houses and barns. Our illustration is from a photograph of a labourer's cottage near Leatherhead, under whose tiles an astonishing number of Swifts nest every year. The tenant informed us that he could hardly ever sleep after daybreak on account of the noise made by the hungry young Swifts. Throughout the British Isles; though there are districts where the bird is not met with.

Materials.—Scarcely any, consisting only of a few straws lined with feathers, and often glued or cemented together with viscid saliva. The Swift will turn to account the old nest of any other bird, provided it is suitably situated, or even lay its eggs amongst a collection of cobwebs and dust.

Eggs.—Two, sometimes three and even four; white, unmarked, and of a narrow elongated shape. Size about 1·0 by ·66 in.

Time.—May and June.

Remarks.—Migratory, arriving in April and May, and leaving in August and September. Note, a harsh scream. Local and other names: Black Martin, Screech Martin, Screamer, Squeaker, Deviling. Gregarious, and very fond of its old nesting haunts. Sits close by.

TEAL.

Description of Parent Birds. — Length about fourteen inches and a half; beak of medium length, fairly straight, and almost black. Irides hazel. Head and upper neck chestnut; a narrow line of buff starts from the base of the bill, goes upward, and passes over the eye and ear-coverts, and onward to the back of the head; a second commences in the front corner of the eye, passes under it, and ends behind the ear-coverts. All the feathers between these two lines are of a rich glossy green; back of lower part of neck, scapulars, and upper part of back, waved or barred with narrow transverse black and white lines; lower part of back shaded with dark brown; wings dark brown, beautifully barred with a patch of glossy green and a line of white; upper tail-coverts nearly black, edged with reddish-brown; tail feathers pointed and brown; lower half of neck, in front, pale purplish-white, spotted with black; breast and belly dusky white; sides and flanks barred with fine wavy lines of black and white; under tail-coverts velvet black; legs, toes, and webs greyish-brown.

The female is much subdued in coloration; her head is light brown, speckled with a darker tint of the same colour; the green spangle on the wing is velvety black; back dark brown, the feathers being edged with a lighter tinge of the same colour; breast and under parts dull white, spotted with dark brown. The male assumes female plumage about the end of July.

Situation and Locality.—On the ground amongst rushes, sedges, heath, and coarse grass, near lakes and small sluggish streams, in mountain swamps, by pools and tarns, and in moss bogs. In nearly all suitable districts throughout the British Isles, perhaps scarcest in the south. Our illustration was procured near a famous mere in Norfolk.

Materials.—Dried sedges, flags, rushes, reeds, and grass, lined with down from the bird's own body.

Eggs.—Eight to fifteen, usually nine or ten; buffish or creamy white, sometimes very faintly tinged with green. Size about 1·7 by 1·35 in. Distinguishable from those of the Garganey only by down tufts which are brown without white tips.

Time.—May.

Remarks. — Resident and partially migratory. Notes: call, *krik;* alarm, *knake.* Local and other names: Common Teal. Sits closely.

TERN, ARCTIC.

Description of Parent Birds.—Length about fifteen inches. Bill, rather long, straight, slender, sharp-pointed, and pinky-red. Irides dark brown. Upper part of head and nape black ; back, wing-coverts, and wings, French grey ; tail-coverts and quills white, with the exception of the two longest feathers, which are grey. The wings are very long, and the tail much forked. Cheeks and chin white ; throat and sides of neck ash grey ; breast, belly, and vent, French grey. Legs, toes, and webs orange ; claws black.

The female is similar to the male.

Situation and Locality.—On the ground, amongst sand, shingle, and on bare rock, near the edge of the water ; on low islands, and at suitable places on mainland shores ; on the Farne Islands, on the Yorkshire side of the mouth of the Humber ; the Scilly Isles ; on the Welsh coast, Lancashire, Cumberland ; generally round the Scottish coast, and at various suitable places in Ireland. Our illustration shows a number of the birds sitting on their eggs at the Farne Islands.

Materials.—None whatever, as a rule, and I am inclined to think that where bits of grass and seaweed are found their presence is accidental.

TEAL.

Eggs.—Two to three, varying in ground colour from pale bluish-green to brownish-buff, blotched and spotted with varying shades of brown and grey. They are found nearer the water's edge, as a rule, are slightly smaller, more boldly marked, and inclining to green in the tinge of their ground colour, than those of the Common Tern. It is, however, often very difficult to distinguish them. Size about 1·55 by 1·1 in.

Time.—May and June.

Remarks.—Migratory, arriving in April and May and departing in September and October. Notes: a prolonged *krr-ee*. Local and other names: none. Gregarious. A light sitter, the whole colony, when visited, rising and fluttering overhead in a noisy throng. The bird is not shy, however, and will alight after a little while and sit on its eggs within fifteen or twenty yards of the intruder.

TERN, COMMON.

Description of Parent Birds.—Length about fourteen and a half inches. Bill rather long, slender, straight, sharp-pointed, and pinky red in colour, except at the tip, which is black. Irides dark brown. Upper part of head and nape black; back ash grey; wings very long, and same colour as the back. Tail much forked and white, except the outer webs of the two longest feathers, which are ash-grey. Chin, throat, breast, belly, vent, and under tail-coverts white, distinguishing the bird from the Arctic Tern, which is grey on its under-parts. Legs, toes, and webs crimson; claws black.

Situation and Locality.—A mere hollow on the ground, amongst shingle, sand, coarse grass and vegetation, on rocks and dried wrack; on small

ARCTIC TERNS ON THEIR NESTS AT THE FARNE ISLANDS.

islands and quiet stretches of shingly beach round
the coasts of the British Isles. Less numerous
round the northern and western coasts and islands
of Scotland than the Arctic Tern, but more numerous
on the southern and western coasts of England.

Materials.—Dry grass, used as a lining to the
slight declivity made or chosen. When the eggs
are laid on bare rocks, sometimes a slight kind of
mat of grass is made. Often there is no kind of
material whatever, even when the eggs are laid
in this situation. Our illustration is from a photograph taken on the Farne Islands.

Eggs. — Two to three. Ground colour light
stone or buff to olive or umber brown, with ash-grey
and light and dark brown spots. Subject to great
variation. Size about 1·7 by 1·15 in. A trifle larger
than those of the Arctic Tern, less boldly marked
and lacking green tinge.

Time.—May and June.

Remarks.—Migratory, arriving in May and departing in August, September, and October. Note, a
sharp, angry *pirre.* Local names: Sea Swallow,
Tarney or Pictarney, Tarrack, Tarret, Rittock,
Rippock, Sporre, Scraye, Pirr, Gull Teaser. Gregarious ; sits lightly, and flies overhead when disturbed, uttering its sharp cry.

TERN, LESSER. *See* TERN, LITTLE.

TERN, LITTLE. *See* LESSER TERN.

Description of Parent Birds.—Length between
eight and nine inches. Bill fairly long, straight, and
orange-coloured, except at the tip, which is black.

COMMON TERNS.

Irides dusky. Forehead white, crown and nape deep black. Back and wings French grey, the first two wing-quills being a trifle darker; the wings are long and narrow. Upper tail-coverts and tail, which is much forked, white. Chin, throat, sides of neck, breast, belly, and vent, clean glossy white. Legs, toes, and membranes orange.

Situation and Locality.—On the ground, on sandy, flat coasts interspersed with banks of shells, gravel, and shingle. Some authorities assert that the bird scrapes a slight hollow for the reception of its eggs, whilst others deny this. Small colonies are still said to breed on the Kentish side of the mouth of the Thames; also on the coasts of Essex, Suffolk, Norfolk, Lincolnshire, Yorkshire, Cumberland, and Lancashire, and in suitable places round the Welsh, Scottish, and Irish coasts. It also breeds in a few suitable inland places, but is decreasing in numbers.

Materials.—None, the eggs harmonising well with their surroundings.

Eggs.—Two to four, generally two or three, varying in ground colour from stone yellow to pale brown, spotted, speckled, and blotched with grey and dark chestnut-brown. Size about 1·25 by ·95 in. Distinguished by smaller size of eggs and also of parent birds.

Time.—June.

Remarks.—Migratory, arriving in May and departing in September or early October. Note, a sharp *pirre*. Local and other names: Lesser Sea Swallow, Lesser Tern. Gregarious. When a colony is visited the birds fly boldly round, uttering their sharp cry, and settle quite close to the intruder.

TERN, ROSEATE.

Description of Parent Birds.—Length about
fifteen and a half inches. Bill rather long, straight,
and sharp pointed; from the tip to the nostrils
it is black, and from thence to the base red.
Irides dark brown. Crown and back of head black;
neck white all round; back and wings ash-grey;
tail much forked, long, and pale ash-grey. Breast,
belly, sides, vent, and under tail-coverts white,
tinged strongly with rose colour. Legs, toes, and
webs red. It may be distinguished from the other
members of the Tern family by its rose-coloured
under-parts and elegant, attenuated form. It has been
mentioned as occupying the same place amongst
Terns as the greyhound does amongst dogs.

The female is very similar to the male.

Situation and Locality.—On the ground, in a
slight hollow, amongst sand and shingle, on low
rocky islands, such as the Farne and Scilly. So
far as the British Isles are concerned, this beautiful
bird, as a nesting species, was supposed to have
become banished. A few pairs tried to re-establish
themselves on the Farne in 1880, but were recognised, and all shot or driven away by a light-house
attendant, despite the law for their protection
being then in force. It is pleasant, however, to be
able to record that, according to the keepers, two
pairs succeeded in rearing their young at the Farne
Islands in the year 1894.

Materials.—Sometimes with, and at others without, a slight lining of bents.

Eggs.—Two to three. "Ground colour creamy
white or buff-brown, blotched and clouded with
bluish-grey and rich brown," according to Mr.

Saunders. Subject to same variations as the Common and Arctic Terns, but a trifle longer than the second. Distinguishable, however, only by the appearance of parent birds. Size about 1·7 by 1·15 in.

Time.—May and June.

Remarks.—Migratory, arriving in May and departing, probably, about the time of the other Terns. Note: *krr-ee.* Local and other names: none. Gregarious, and not a close sitter.

TERN, SANDWICH.

Description of Parent Birds.— Length about fifteen inches; bill rather long, straight, slender, and pointed; black in colour, except at the tip, which is primrose-yellow. Irides hazel. Crown of the head and nape black; the feathers at the back of the head are rather long and form a loose, pointed plume; back pearl-grey; wings very long, pearl-grey in colour, except the longest quill-feathers, which are rather darker; tail white and much forked; chin, throat, breast, belly, and vent white, sometimes tinged with salmon pink; legs, toes, membranes, and claws black.

The female is very similar to the male.

Situation and Locality.—On the ground, in a slight hollow in the sand or pebbles on low rocky islands. The principal colony in this country is on the Farne Islands. Our illustration is from a photograph taken on a slight ridge of sand, measuring about twenty by seven yards. The keepers informed us that in 1892 they counted 210 Sandwich Terns' nests on the same ridge. When our boat neared the Tern Island (Knoxes) the birds

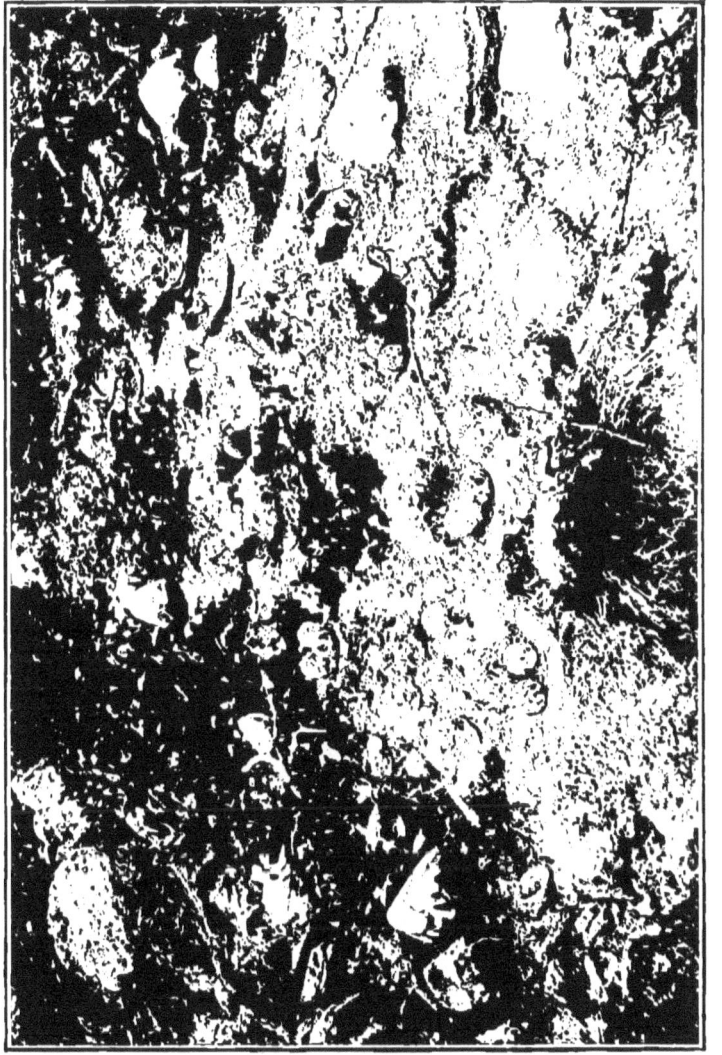

SANDWICH TERNS' NESTS AT FARNE ISLANDS

(Sandwich, Arctic, and Common) rose in a perfect cloud, and whirled round and round, for all the world like a thick shower of snow played upon by fickle gusts of wind. Small colonies breed at various points on the Scottish and Irish coasts, also at Walney Island, off Lancashire, and at Ravenglass, in Cumberland.

Materials.—Generally none; sometimes, a few bits of dead herbage, and occasionally even a liberal supply of dry grass is used.

Eggs.—Two to three, usually the former number; varying very much in colour, from creamy-white to buffy stone-colour, spotted, blotched, and clouded and scrolled with grey, deep rich brown, and chestnut. I saw one on the Farne of a uniform dark brown, although its fellow egg was of the normal coloration. Size about 2·1 by 1·4 in. Size of parent birds and eggs prevent confusion with those of any other Tern.

Time.—May and June.

Remarks.—Migratory, arriving in April and May and departing in August. Note, a hoarse and grating kind of *kirkitt* or *kirhit*. Local and other names: Great Tern. Gregarious, sits lightly, and flies round the intruder, uttering its hoarse cry.

THICKNEE. *See* CURLEW, STONE.

THRUSH, COMMON. *See* THRUSH, SONG.

THRUSH, MISSEL. *Also* MISTLETOE THRUSH.

Description of Parent Birds. — Length about eleven inches; bill moderately long, slightly curved downwards, and dark brown, with a yellowish tinge at the base of the lower mandible. Irides hazel. Top of the head, nape, and back, light brown; rump yellowish-brown; wings dark brown, the coverts being tipped and the quills edged with wood-brown; the tail is darkish brown, two or three of the outside feathers being tipped with greyish-white; cheeks, chin, throat, breast, and under-parts straw colour or yellowish-white, lightest on the belly and vent; the throat and breast are marked with triangular spots of blackish-brown, and the belly with roundish ones of the same colour; legs and toes light brown; claws brownish-black.

The female is similar to the male. Easily distinguished from Song Thrush by its larger size.

Situation and Locality.—Generally near the top of a tree, where the trunk ends and two or three strong branches spring from it, or on a strong bough close to the trunk, at varying heights from the ground, but seldom or never in the position of the Song Thrush. In woods, plantations, parks, and tree-fringed streams all over the United Kingdom. I have met with its nest much oftener in the north of England than either in the south or east, and have noted its partiality for a lichen-covered ash tree. It is a brave bird, and I have seen it sitting on its eggs when one side of the tree has been plastered white with wind-driven snow. Our illustration was procured high up amongst the Westmoreland hills.

Materials. — Slender twigs, grass stems, wool, moss, and mud, with an inner lining of fine dry grass. The wool often hangs down from the sides of the nest in long conspicuous rags.

Eggs.—Four to six, according to some authorities. Mr. Dixon, however, says never more than four.

Mr. Seebohm says they very rarely exceed four, and in but very few cases are less. Messrs. Dresser and Sharpe say the number is generally five, sometimes four; Waterton says generally five; Macgillivray, usually four, or from three to five. I have certainly heard of more than four; but although I have taken and examined a goodly number of nests, I personally never saw more. They vary in colour, some being greyish-green with underlying markings of grey, and blotches and spots of reddish-brown. Others are reddish-grey in ground colour, with brownish-red markings, which vary in size and distribution. Size about 1·3 by ·88 in.

Time.—February, March, April, May, June, and July. I have found them in every month but the last.

Remarks.—Resident, but subject to southern movement in winter. Song loud and defiant, but not considered of much value by bird-fanciers, as it is said to be melancholy and made up of five or six broken strains; alarm note, a jarring kind of scream. Local and other names : Holm Thrush, Storm Cock, Holm Screech, Mistletoe Thrush, Missel Bird, Bell Throstle, Screech Thrush. Sits pretty closely, and makes a great deal of demonstration when disturbed.

THRUSH, SONG. *Also* Thrush *or* Common Thrush.

Description of Parent Birds.—Length about eight and a half inches. Bill of medium length, nearly straight, and dusky. Irides hazel. Head, nape, back, wings, rump, tail-coverts, and quills yellowish-brown, spotted with darker brown on the sides

MISSEL THRUSH

of the head, and edged with a lighter tinge on the
wing-quills. Throat, breast, and under-parts pale
tawny yellow; lighter on the vent and under tail-
coverts. The space from the throat to the thighs
is studded with arrowhead-like spots. Legs and
toes pale brown; claws darker.

The female is smaller than the male, and the
spots on her breast are larger and the ground colour
lighter.

Situation and Locality.—In evergreens, especially
early in the spring, hedgerows, bushes, in ivy growing
against walls and trees, in holes and on "throughs"
of dry walls; on ledges of rock, on beams and in
holes of barns, and sometimes quite on the ground;
in woods, plantations, on commons, hedges, trees
and bushes growing by the side of brooks. Our
illustration is from a photograph taken early in
the spring. Throughout the British Isles, with
few exceptions, and those where no cover is afforded.

Materials.—Twigs, coarse grass, moss, and clay,
with an inner lining of cow-dung or mud; sometimes
thickly studded with bits of rotten wood.

Eggs.—Four to six, of a beautiful deep greenish-
blue, spotted with black. The spots sometimes
describe a well-defined ring round the larger end,
at others they are sparingly scattered over the
egg, and in rare cases are absent altogether. Very
variable in size. Average measurements about 1·05
by ·8 in.

Time.—February, March, April, May, June, and
July; sometimes as late as August, and even October.
I have seen the bird bravely covering her eggs when
the ground has been thickly mantled in snow.

Remarks.—Resident, subject to local movement,
and partially migratory. Notes: call, *sik, sik, sik, sik,
siki, tsak, tsak.* The song of the cock is well known

SONG THRUSH

and highly esteemed. In North Yorkshire it is verbalised as *Pay thy debt, pay thy debt, skitting Mick, skitting Mick.* Local and other names : Throstle, Mavis. Sits closely, and protests loudly against molestation.

TIT, BEARDED.

Description of Parent Birds.—Length just over six inches, the tail forming about half of this ; bill short, upper mandible slightly curved downwards, and yellow. Irides yellow. Crown bluish-grey ; nape, shoulders, back, and rump, golden brown ; wings black and greyish, the feathers being bordered and tipped with white and deep rusty red ; upper tail-coverts and tail, which is graduated and wedge-shaped, deep rufous brown, some of the outer feathers being tipped and edged with white and greyish white. A black patch extends from the base of the bill to half way over the eye and, passing downwards, ends in a tuft of elongated, tapering black feathers growing from the side of the chin and throat very like a moustache. Centre of chin and throat dirty white : breast flesh-colour : belly and vent like the back, but brighter ; under tail-coverts black ; legs, toes, and claws black.

The female has the crown dull rusty brown ; tufts on the sides of the chin pale brownish-fawn ; chin and throat mixed with light brown ; breast of a lighter tinge than in the male ; and under tail-coverts pale golden-brown.

Situation and Locality.—Near the ground, in a tuft of coarse grass, nettles, or sedge ; sometimes amongst broken-down reeds, but never suspended between stems of any kind. It is well hidden, but now, alas! only known for certain to be built

amongst the extensive reed-beds and marshes round
the Norfolk Broads, from whence it is likely to
become banished at no distant date. Indeed, it is
rumoured that the severe winter of 1894-5 has
annihilated it.

Materials.—Dead aquatic vegetation, such as
leaves of reeds and blades of sedges, lined with
fine grass and seed-down. It is cup-shaped.

Eggs.—Four to seven, white, faintly tinged with
cream colour, and marked with small specks, short
irregular streaks and splashes of dark brown. Size
about ·7 by ·56 in. Distinguished by situation, size
of eggs, and streaky markings.

Time.—March, April, May, June, and July.

Remarks.—Resident, but subject to local movement. Notes shrill and musical when alarmed.
They also utter a clear silvery note before alighting
which sounds something like *ting, ting, ting.*
Local and other names: Reed Pheasant. Sits
closely, and is difficult to see on account of its
habit of hiding amongst reeds.

TIT, BLUE.

Description of Parent Birds.—Length from four
to four and a half inches. Bill short, strong and
dusky. Irides dark brown. Crown clear blue,
under which runs a band of white on either side.
From the base of the beak, through the eyes, passes
a bluish-black line. Cheeks white; a broadish circle
of dusky blue runs round from the back of the head
to the throat, where it becomes almost black. Back
and rump lemonish-green; wings and tail blue,
the former marked with white on the coverts and
tertials. Chin and throat blue-black; breast, belly,

and under-parts yellow. Legs, toes, and claws dull leaden-blue.

The female is less brilliant and distinctive in her coloration.

Situation and Locality.—In holes in trees, walls, banks, and often in such queer places as disused pumps, letter-boxes, stone bottles, flower-pots, boxes and cocoanuts, hung in trees for its accommodation. Our illustration is from a photograph of a nest in a hollow fruit tree. The entrance to the nest was in the centre of a decayed branch which had been sawn or broken off close to the trunk, at the place where the dark excrescence-like growth appears, the hole through which the eggs are to be seen being cut artificially through the wood, so as to show its exact position. In barns, stables, cottages, orchards, gardens, woods, and cultivated districts generally, throughout the United Kingdom, with exception of the islands lying to the West and North of Scotland.

Materials.—Grass, moss, hair, and wool; sometimes a few soft leaves woven together, with an inner lining of feathers. I have met with specimens containing few or none of the last.

Eggs.—Six to nine, sometimes as many as eleven or twelve, white, spotted with light red or red-brown, sometimes evenly distributed, at others most numerous at the larger end. Size about ·6 by ·46 in. A sight of parent birds only will definitely settle identification.

Time.—April, May, and June.

Remarks. — Resident. Note: a peculiar *twe-twe*. Local and other names: Tomtit, Blue Tomtit, Billy Biter or Willow Biter, Blue Bonnet, Blue Cap, Blue Mope, Hickwall, Nun, Titmal. A close sitter, hissing like a snake when disturbed.

BLUE TIT.

TIT, COAL.

Description of Parent Birds.—Length about four
and a half inches. Bill short, straight, pointed, and
black. Irides hazel. Head, neck, and upper part of
breast black ; cheeks and nape white. Back, wing-
coverts, rump, and tail greyish-blue, with a buffish
tinge on the rump. Wing-quills brownish-grey,
bordered with green. Lower breast dull white ;
belly, flanks, vent, and under tail-coverts fawn
colour, slightly tinged with green. Legs, toes, and
claws black.

The female closely resembles the male. This
bird is easily distinguished from the Marsh Tit
by means of the white patch on the nape of its
neck.

Situation and Locality.—In holes, from three or
four to sixteen or eighteen inches deep, in trees,
walls, and banks ; those in the last-named situations
having originally belonged to rats, mice, or moles.
The bird will, however, enlarge any selected hole for
its accommodation, if necessary. It may be found
in orchards, spinneys, coppices, woods, and plantations throughout England, Wales, Scotland, and
Ireland.

Materials.—Dry grass, moss, wool, and hair,
lined liberally with feathers.

Eggs.—Five to ten, generally seven or eight,
white, spotted and freckled with light red, the
markings being generally most numerous at the
larger end. They bear a very close resemblance
to those of other members of the Tit family, but
a sight of the parent birds will readily distinguish
them. Size about ·62 by ·47 in.

Time.—April, May, and June.

Remarks — Resident. Note: a harsh, shrill *che-chee, che-chee.* Local and other names: Colemouse, Coal Titmouse, Coalhead. Sits closely, and hisses when molested.

TIT, CRESTED.

Description of Parent Birds.- Length about four and a half inches; bill short, straight, and almost black. Irides hazel. The feathers on the top of the head, especially behind, are lengthened, and form a conspicuous crest; these feathers are dull black in colour, tipped with light grey; back, wings, rump, and tail, brown, the quills being somewhat darker. A black streak runs from the base of the beak to the eye, and passes onward between the base of the crest and the ear-coverts to the nape, from whence a broader black, curving line descends behind the cheeks and ends abruptly on the sides of the neck. Cheeks white, spotted with black. The black line just described is followed beneath by a broader band of white, which in turn gives place to a narrower curving black line, descending from the back of the head, and, passing in front of the point or shoulder of the wing, joins the black on the upper breast; chin, throat, and upper breast black; lower breast, belly, and flanks dull white, suffused with buff on the sides; under tail-coverts dull buff; legs, toes, and claws lead-grey.

The female resembles the male, except that her crest is shorter, and the black on her chin, throat, and upper breast occupies less space.

Situation and Locality.—Holes in the trunks, branches, and old stumps of trees. The hole is situated from a few inches to ten or twelve feet

U

from the ground, and is sometimes dug by the bird's own exertions. On the Continent it patronises the deserted nests of Crows and old squirrel dreys. It breeds in old pine, fir, and oak forests in Ross, Banff, Perth, Inverness, and possibly one or two other favourite localities in Scotland only.

Materials.—Dead grass, moss, feathers, and down of hares and rabbits.

Eggs.—Five. Professor Newton says that they do not seem to exceed five in number. Mr. Dixon gives the figures as from five to eight, and Mr. Morris from seven to ten. Possibly the last is the result of some Continental information, as it is said the bird lays from eight to ten eggs there. White, spotted, blotched, and speckled with red of varying shades; the spots are generally most numerous round the larger end. Size about ·65 by ·51 in. A sight of the parent birds and locality of the nest are the only safe means of identification.

Time.—April, May, and June.

Remarks.—Resident and very local. Notes, *si, si, si,* followed by a spluttering note like *ptur, re, re, re, ree,* according to Mr. Seebohm. Local and other names: Crested Titmouse. Sits closely.

TIT, GREAT.

Description of Parent Birds.—Length about five and three-quarter inches; bill of medium length, nearly straight, and black. Irides dusky. Head black; cheeks white; nape greenish-yellow, surrounding a whitish spot; back olive green; rump bluish-grey; wing-coverts bluish, the larger being tipped with white; quills dusky, edged with light greenish-blue; tail quills dusky, the outer feathers

on each side being edged with white; breast and
belly yellow, tinged with green and divided longitudinally by a broad black stripe, which commences
at the chin, and passing down the throat, is joined
by the black on the sides of the neck, and extends
to the vent, which is white on either side; lower
tail-coverts whitish; legs, toes, and claws bluish-
grey.

The female is not so distinctive in coloration;
the black stripe on her under-parts is not so wide
or pronounced, and vanishes about the middle of
the belly. She is also said by some ornithologists
to be smaller.

Situation and Locality.—In holes in trees, walls,
posts, banks, and buildings. The bird will excavate a
hole for itself in a rotten tree sometimes. It also
occasionally fixes its abode beneath the nest of a
Crow, Rook, or Magpie. The hollow tree shown in
our illustration is resorted to as a nesting site nearly
every year by this bird, whose little abode has
been recorded in a drinking-cup and a plant-pot.
In orchards, gardens, yards, well-timbered commons,
woods, plantations, parks, and other wooded places
throughout the United Kingdom with few exceptions.

Materials.—Dry grass, moss, hair, wool, and
feathers, somewhat carelessly put together. Sometimes no materials at all. Montagu and Morris
supposed that where materials were dispensed with
altogether, the bird had had her first eggs and nest
taken, and had not had time for more nest-building.
This theory, I am pleased to be able to say from
my own experience, is perfectly sound. A year
or two back I took a nest and eggs, quite fresh,
from a hollow apple-tree, and passing by the old
orchard a little while after I found that the bird

had laid a second clutch of eggs on the rotten powdered wood at the bottom of the hole.

Eggs.—Five to twelve, generally seven or eight; white in ground colour, spotted and freckled with pale red. The markings vary in size, number, intensity, and position. Sometimes they are generally distributed, at others they form a zone at the larger end of the egg. They are larger than the eggs of any other Tit, but are likely to be confused with those of the Nuthatch if care is not taken. Size about ·7 by ·53 in.

Time.—April, May, and June.

Remarks.—Resident. Notes, a clear, pealing *pinker, pinker*, repeated several times. Local and other names: Oxeye, Pickcheese, Great Black-headed Tomtit, Blackcap, Great Titmouse, Bee-biter, Tom Collier, "Sit-ye-down." Sits closely, and hisses when disturbed.

TIT, LONG-TAILED.

Description of Parent Birds.—Length about five and a half inches, of which the bird's abnormally long tail forms a considerable part; bill very short, curved above and below, and black. Irides hazel. Forehead and crown white. A black streak of variable width commences near the base of the bill and, passing over the eye and ear-coverts, meets on the back of the neck, and descending forms a kind of triangle, the lowest point of which reaches the middle of the back. The scapulars and rump are suffused with a dull purplish-red; wing-coverts and quills black, the inner ones of the latter edged with white; upper tail-coverts and six centre quills black, the remainder of the feathers on either side of the tail being more or less white. The feathers

HOLLOW APPLE TREE IN WHICH A PAIR OF GREAT
TITS NEST ANNUALLY.

of the tail are much graduated in length, those in
the centre being about an inch and three-quarters
longer than those on the sides. Cheeks and ear-
coverts white, the latter mixed with grey. Chin,
throat, breast, belly, and under-parts greyish-white,
tinged with purplish-red on the sides, flanks, vent,
and under tail-coverts; legs, toes, and claws almost
black.

The female is somewhat similar, with the exception of her head, which has more black upon it;
however, both sexes are subject to variation in the
intensity of coloration.

Situation and Locality.—In whitethorn hedges,
sloe, gorse, and wild rose bushes. I have a specimen
taken from a bramble bush, which is very similar
to a bottle. Our illustration is from a photograph of a nest situated in a low, thick hedge
near London. It is found in nearly all suitable
localities throughout the British Isles.

Materials.—Moss, lichens, wool, spiders' webs,
cunningly felted together, and skilfully formed into
an oval-shaped nest, which is plentifully lined with
feathers and securely fastened to its surroundings.

Eggs.—Seven to ten; as many as twenty have
been found, but such a number was undoubtedly
the production of two hens. White or rosy-white
until blown, by reason of the yolk showing through
the delicate shell, with very small, faint red or
reddish-brown spots, sometimes collected round the
larger end, at others sparingly scattered over the
entire surface; occasionally without spots altogether.
Size about ·57 by ·44 in. The smaller number of
spots, the character and situation of the nest, and
the appearance of the parent birds readily identify
the eggs of this Tit.

Time.—March, April, May, and June.

LONG-TAILED TIT.

Remarks.—Resident. Notes, a sharp chirrup or twitter, varied by a lower and hoarser note. Local and other names: Mufflin, Poke Pudding, Longpod, Bottle Tit, Oven-builder, Caper Long-tail, Long-tail Pie, Long-tailed Capon, Bottle Tom, Mum Ruffin, Long-tailed Mag, Huckmuck (a name also applied to the Willow Warbler), Longtailed Mufflin, etc. Sits closely, with the tip of her tail protruding from the hole.

TIT, MARSH.

Description of Parent Birds.—Length about four and a half inches. Bill short, straight, sharp-pointed, and black. Irides dark hazel. Forehead, crown, and nape deep black. Back, wing-coverts, and upper tail-coverts ashy brown, mixed with a greenish tint. Wing and tail-quills greyish-brown, with edges of a lighter tinge. Cheeks dirty white; chin black; throat and breast dull greyish-white; belly and vent of the same colour, tinged with brown. Legs, toes, and claws bluish-black.

The female is similar in appearance to the male. The bird may be easily distinguished from the Coal Tit by its having no white on the nape or wing-coverts.

Situation and Locality.—Holes in trees, preferably pollards, gate-posts, walls, and banks, at no great height from the ground. Instances are on record of rabbit-burrows and rat-holes doing duty as nesting sites. In orchards, woods, by the side of sluggish rivers, and in hedgerows of cultivated districts. It is met with in most parts of England, suitable to its habits. Scotland and Ireland can both claim it, but it is somewhat scarce, especially in the northern parts of both countries.

Materials.—Moss and fine dried grass, lined with wool, feathers, hair, rabbits' down, ripe catkins of the willow, the whole being compactly knitted together, and tightly wedged into the situation chosen for their reception.

Eggs.—Six to ten, white, spotted with reddish-brown, more thickly at the larger end. The spots are variable in size, number, and distribution. They very closely resemble those of the Creeper, Blue Tit, Coal Tit, Nuthatch, and Great Tit, although, as a rule, they are somewhat smaller than the last. Size about ·63 by ·49 in.

Time.—April, May, and June.

Remarks.—Resident. Notes: all, *chee-chee* or *peh, peh*, uttered quickly, and several times in succession, and a kind of whistle, made use of only in the spring, according to Montagu. Local and other names: Black Cap, Little Black-headed Tomtit, Willow Biter, Coalhead. Sits closely, and hisses and bites when disturbed.

TWITE. *Also* MOUNTAIN LINNET.

Description of Parent Birds.—Length about five and a quarter inches; bill short, broad at the base, and pale yellowish flesh colour. Irides hazel. Crown, neck, back, and upper tail-coverts dark brown, the feathers being edged with light rufous-brown; rump purplish-red; wing and tail quills very dark brown, more or less edged on the outer webs with white. The feathers round the base of the beak and below the eyes tile-red; sides of the head dark brown, edged with a lighter tinge; chin and throat rufous, lighter on the breast and sides, which are speckled with brown; belly nearly white; vent tinged with brown; under tail-coverts almost white; legs, toes, and claws dusky; tail slightly forked.

The female is lighter coloured on her upper parts, and lacks the red on her rump.

Situation and Locality.—Amongst tall heather, ling, brushwood, and furze. Sometimes quite on the ground against the side of a bank or by a stone in moorland districts in the North of England, Scotland and the surrounding islands, and Ireland. Our illustration is from a photograph taken on a small island near Oban, where we found several nests.

Materials.—Twigs, fibrous roots, grass stalks and blades, moss, and wool, with an inner lining of feathers, hair, or down.

Eggs.—Four to seven, generally five or six: very similar, indeed, to those of the Linnet. Pale bluish-green, spotted and streaked with reddish-brown and dark brown; sometimes streaked with the lighter reddish tinge. Some authorities say that they are a little more streaked, and that the light red markings are less frequent than in those of the Linnet. The markings are generally most numerous on the larger end of the egg. Size about ·69 by ·5 in. Easily distinguished by the appearance of the parent birds.

Time.—May and June.

Remarks.—Resident in its breeding haunts, but a winter visitor to the more southern portions of England. Notes, *twite;* the cock has a pleasing little song. Local and other names: Mountain Linnet, Twite Finch, Heather Lintie. Sits closely.

TWITE

WAGTAIL, BLUE-HEADED.

Description of Parent Birds.—Length about six and a half inches; bill fairly long, slender, straight, and black. Irides dullish brown. Crown and nape bluish-grey; scapulars, back, and upper tail-coverts greenish-olive, suffused with yellow; wing-coverts and primaries dark brown, the former, as well as the tertials, bordered with greyish-yellow. The tail is black in the centre and white on the outer edges. Over the eye and ear-coverts is a white streak, also one of shorter dimensions under the eye; ear-coverts bluish-grey; chin and cheeks white; throat, breast, belly, vent, and under tail-coverts golden yellow; legs, toes, and claws black.

The female is somewhat smaller and less brilliant and distinctive in coloration. The bird may be distinguished from the Yellow Wagtail, which it closely resembles, by its bluish-grey head and the white streak over the eye and ear-coverts.

Situation and Locality.—On the ground, amongst meadow grass, on hedgerow banks, amongst the exposed roots of trees, in pastures, grass meadows, and cornfields, according to Continental observations.

Materials.—Dead grass, moss, and fibrous roots, lined with horsehair.

Eggs.—Four to six, usually five; quite similar to those of the Yellow Wagtail; greyish-white, suffused, mottled, or spotted with varying shades of brown; sometimes marbled with a few fine lines of dark brown. Size about ·78 by ·56 in.

Time.—May and June, according to Messrs. Dixon and Miller Christy; but these months are either based upon Continental observations or deductions from the laying season of the Yellow Wagtail.

Remarks.—About forty specimens of the Blue-headed Wagtail have been procured in this country, shot in January, April, June, and October. Its nest has only been met with on one or two occasions at Gateshead, but it is thought that from its close similarity to the Yellow Wagtail it has often been overlooked. Call note, *chit-up.* Local and other names : Grey-headed Wagtail. There is, so far as I can gather, no precise information forthcoming as to whether the bird is a close sitter or not.

WAGTAIL, GREY.

Description of Parent Birds.—Length about seven and three-quarter inches, nearly half of which is accounted for by its unusually long tail. Bill of medium length, nearly straight, and dusky brown. Irides dark hazel. Crown and sides of head bluish-grey; a narrow white streak runs over the eye and ear-coverts. Back of neck, back, scapulars, and rump bluish-grey; wing-coverts black, or very nearly so, tipped with buffish-white; quills black, some of the inner ones edged on the outer webs with yellowish-white, and liberally marked on the inner, towards the base, with white. Upper tail-coverts greenish-yellow; tail black, yellowish on the edges of the centre feathers towards the base; the two outside quills on either side white, with the exception of a narrow black line on the outer web of the second feather. Chin and throat black, separated from the sides of the head and neck by a white line; breast, belly, and under-parts bright yellow. Legs, toes, and claws pale brown.

Situation and Locality.—On shelves of rock, in crevices, in rough, rocky, and uneven banks, holes in stone walls, behind or under large stones, rarely far away from water. It is very local, and, like the Dipper, seems to lay claim to a certain length of stream. I know two waterfalls on moorland becks in the North of England where a Dipper and a Grey Wagtail nest almost yearly within a few yards of each other. I have known the bird on one occasion become foster-parent to a young Cuckoo. It breeds in the western and northern counties of England; in Wales, Scotland, and in parts of Ireland. Our illustration was procured in Westmoreland.

Materials.—Rootlets, grass, and moss, lined with horse and cow hair: sometimes a few feathers.

Eggs.—Four to five, occasionally six, of a greyish-white ground colour, spotted and speckled with pale brown. Sometimes the ground colour is buffish and the markings creamy-brown. Occasionally a few streaks of dark brown are present. Size about ·75 by ·56 in. Much like the eggs of the Yellow and Blue Headed Wagtails, also Sedge Warbler, but easily identified by locality, situation, and a sight of the parent birds.

Time.—April, May, and June.

Remarks.—Resident, but subject to local migration. Notes: *sziszi* or *zisy*, sharply uttered. Local and other names: Dun Wagtail, Nanny Washtail, Grey Wagster. Sits closely, and when disturbed hovers round with her mate, uttering a note of alarm.

GREY WAGTAIL.

WAGTAIL, PIED.

Description of Parent Birds.—Length about seven and a half inches, several of which are accounted for by the somewhat abnormally long tail. Bill moderately long, nearly straight, slender, and black. Irides dusky. Forehead, sides of head, round the eyes, and a portion of the sides of the neck, white. Latter half of crown, nape, back, and upper tail-coverts black ; wing-coverts black, edged and tipped with white ; quills black, some of them bordered with greyish-white and white ; tail-quills black, except the two outside feathers, which are nearly all white. Chin, throat, sides, and flanks black ; breast, belly, and under-parts white. Legs, toes, and claws black.

The female is somewhat smaller, and dusky grey on the back, where the male is black.

Situation and Locality.—In ivy, growing against walls and trees, in holes in dry walls, bridges, niches of rock, on ledges, and tufts of grass growing from crevices of rock ; in faggot, hay, and brick stacks, and numerous other situations, generally near fresh water, throughout the British Isles. It has been recorded in such curious situations as a potato top, and under a railway switch. Our illustration is from a photograph of a nest situated inside a reed-thatched Norfolk boat-house.

Materials.—Dry grass, roots, and moss ; sometimes a few dead leaves or fern-fronds, with an inner lining of wool, feathers, horsehair, cowhair, and rabbit down. The materials vary both in quantity and character, according to situation.

Eggs.—Four to six, greyish-white, thickly speckled with ash-grey or light brown. They vary a good

PIED WAGTAIL.

V

deal, according to the tint of the ground colour,
size of the spots, and their colour. Those in the
nest represented were of a bluish-white, marked
all over with small ash-grey spots. Size about ·8
by ·6 in. Indistinguishable from those of the White
Wagtail or certain varieties of the House Sparrow,
except by the parent birds in the former case and
the nest and its position in the latter.

Time.—March, April, May, and June.

Remarks.—Resident, but with a southern winter
movement. Notes : *chiz-zit, chiz-zit.* Local and
other names : Dishwasher, Black and White Wagtail, Washtail, Nanny Washtail, Wagster, Water
Wagtail, Washerwoman. Sits rather closely.

WAGTAIL, WHITE.

Description of Parent Birds.—Length nearly eight
inches. Bill of medium length, straight, and black.
Crown and nape black ; back, scapulars, and upper
tail-coverts French or light ash-grey. Wings
brownish-black, each feather having a broad outer
margin of greyish-white. Tail-quills black, with
the exception of the two centre feathers, which are
margined with white, and the two outer feathers
on each side, which are white, with black inner
webs. The front and sides of the head, together
with a patch on either side of the neck, are white.
Chin, throat, and upper part of the breast, black.
Lower breast, belly, vent, and under tail-coverts
white. Legs, toes, and claws black.

The female is less distinctive in coloration. Her
forehead and cheeks are not so pure a white ;
throat mottled with white ; black on back of head

occupies less space, and her back is tinged with olive.

Situation and Locality.—Similar in all respects to those of the Pied Wagtail.

Materials.—The same as are employed by the bird just named.

Eggs.—Five to seven, of a wider colour variation than those of the Pied Wagtail, according to Mr. Dixon. The ground colour varies from pure white to bluish-white, speckled all over with different shades of grey and brown; sometimes a few hair-like lines occur at the larger end. The markings vary, both in regard to size and distribution, and there can hardly be any safe means of identification apart from the difference in the parent birds. Size about ·8 by ·6 in.

Time.—April, May, and June.

Remarks.—Migratory, but little is known as to its comings and goings. Although a common bird on the Continent, only a few well-authenticated instances of its breeding in the British Isles are on record, and those in the southern counties of England. It is, however, thought that it may often have been overlooked from the fact that its general appearance and eggs are so similar to those of the Pied Wagtail, to all except the practical and experienced ornithologist. Notes: call, *chiz-zit.* Local and other names : Grey and White Wagtail. Said to sit pretty closely. The male differs from that of the Pied Wagtail in being grey on its upper-parts below the nape instead of black, but the females of the two species only differ in that of the White Wagtail being "pearl-grey or very light ash-grey tinged with olive," and that of the Pied Wagtail being "lead-grey mottled with darker feathers " on those parts.

WAGTAIL, YELLOW.

Description of Parent Birds.—Length about six and a half inches. Bill moderately long, straight, slender, and black. Irides hazel. Crown, nape, back, and scapulars light olive. Wing-coverts and primaries darkish-brown, the first-named being tipped, and the tertials bordered and tipped with greyish-yellow. Upper tail-coverts olive ; tail-quills brownish-black, with the exception of the two outer feathers, which are white, streaked with black on the inner web. Over the eye and ear-coverts is a line of golden-yellow. Chin, throat, breast, belly, and vent a bright golden yellow. Legs, toes, and claws black.

The female is much less handsome, her head and back being darker, and the yellow of her breast and under-parts not nearly so bright.

Situation and Locality.—On the ground, in the shelter of a tuft of grass, heather, or coarse herbage; sometimes behind the long grass of an overhanging bank, well hidden. I know several places in the North of England where pairs breed year after year with unbroken regularity. In grass meadows, pastures, commons, and other suitable places, pretty generally throughout England, except Cornwall and Devonshire. It is much more numerous, according to my observations, in the north than in the south and east ; also in the south of Scotland, and, to a very limited extent, in Ireland. The bird is very wary, and the nest difficult to find. I have watched a pair for three or four hours through my binoculars, and when able to locate the nest pretty closely have still failed to find it.

Materials.—Dead grass, fibrous roots, and moss,

with an inner lining of horse or cow hair, feathers, or down.

Eggs.—Four to six, generally five, ground colour greyish-white, mottled and spotted with varying shades of brown ; sometimes marbled with blackish-brown at the larger end. The markings are thickly distributed over the surface of the egg. They are very similar to those of the Blue-headed Wagtail, Pied Wagtail, and Sedge Warbler. The difference pointed out in regard to the plumage of the first in describing it, and the situations of the two latter ought to prevent confusion. Size about ·78 by ·58 in.

Time.—Some very good authorities say April, but I have never met with a nest so early. May, June, and July.

Remarks.—Migratory, arriving in March or April, and leaving in September. Notes : *tzee-tzee, sipp-sipp.* Local and other names : Cowbird, Ray's Wagtail, Yellow Wagster. Sits lightly, and, although by no means shy, is very wary.

WARBLER, DARTFORD.

Description of Parent Birds. Length about five inches, nearly half of which is accounted for by the bird's exceptionally long tail; bill fairly long, slightly curved downward, and blackish, with the exception of the base of the lower mandible and along the edges of the upper, which are orange. Irides light or dark red, according to age. Head, neck, back, and upper tail-coverts greyish-black ; wings blackish-brown, the quill-feathers being bordered with a lighter tinge ; tail blackish-brown, the external feathers being broadly tipped with grey ;

chin, throat, breast, and sides chestnut-brown; belly white; under tail-coverts slate grey; legs and toes pale reddish-brown; claws darker.

The female resembles the male, except that she is more tinged with brown on her upper- and lighter on her under-parts. The chestnut-brown does not, however, extend so far down the breast.

The Dartford Warbler has the power of partly erecting the feathers on the head, so as to form a kind of crest.

Situation and Locality.—In the lower parts of thick furze bushes; very locally and sparingly on commons and other places covered by furze bushes, principally in the counties along the south coast of England. It was at one time not supposed to nest north of the Thames, but Mr. Dixon has proved that it does so as far north even as Yorkshire. It is not found either in Scotland or Ireland.

Materials.—Small and slender branches of furze, grass stalks, bits of moss and wool, with an inner lining of fine grass, and sometimes a few horsehairs. It is somewhat slight of build, and has been likened to that of the Whitethroat.

Eggs.—Four to five, greenish or buffish-white in ground colour, speckled all over with dark olive-brown, and underlying markings of grey, which generally become more dense at the larger end, and form a kind of zone. There is very little difference indeed between the eggs of this bird and those of the Whitethroat, except that the markings are more conspicuous. Size about ·68 by ·5 in.

Time.—April, May, and June.

Remarks.—Resident. Notes: *pit-et-chou-cha-ch-cha.* Local and other names: Furze Wren. Sits close, and slips away quietly.

WARBLER, GARDEN.

Description of Parent Birds.—Length about six inches; bill fairly long, straight, strong, and dark brown in colour. Irides hazel. Head, neck, back, wings, and tail uniform light brown, slightly tinged with olive; chin, throat, breast, belly, vent, and under tail-coverts dull brownish-white, dark on the throat and breast, and light on the belly; legs, toes, and claws purple-brown.

The female is similar to the male in appearance.

Situation and Locality.—Generally a few feet from the ground in thorn bushes, briars, brambles, gooseberry bushes, nettles, and peas. Sometimes lower down in coarse grass and taller wild plants. In woods, clumps of trees growing beside streams, shrubberies, thick hedges, orchards, and gardens, sparingly, in suitable localities nearly all over England. It also breeds in one or two parts of Wales, in the southern parts of Scotland, and in different parts of Ireland.

Materials.—Straws, blades of grass, fibrous roots, sometimes a little wool or moss, and lined with horsehair. It is a somewhat flimsy structure.

Eggs.—Four to six, generally four or five, varying in ground colour from white to greenish-white or yellowish stone-grey, blotched, spotted, and clouded with brown of various shades; deep olive, with underlying markings of ash-grey. The markings are variously distributed, occasionally being congregated at the larger end. Some specimens are marbled with brown. Size about ·77 by ·6 in. Often indistinguishable from those of the Blackcap, except by a sight of the parent birds.

Time.—May and June.

Remarks. — Migratory, arriving in April and May, and departing in September or October. Notes, song, deep, harmonious, and mellow; call, *tee.* Local and other names : Pettychaps, Fauvette, Greater Pettychaps, Fig Bird. Sits closely.

WARBLER, GRASSHOPPER.

Description of Parent Birds.—Length about five and a half inches; bill of medium length, straight, strong, and brown in colour. Irides brown. Crown, nape, back of neck, back, and wings olive-brown, the centre of each feather being of a darker tinge; tail rather long, much rounded at the tip, and brown, barred with a paler tinge of the same colour; chin, throat, breast, and all under-parts pale brown, darker on the flanks. The neck and breast are spotted with darkish brown; legs and toes pale brown; claws light horn colour.

The female is very similar to the male, but is said to lack breast spots.

Situation and Locality.—On or near the ground under furze and other small bushes, in tufts of tall rank grass growing at the foot of hedgerows, and similar situations affording plenty of cover. In woods, on commons, fens, clumps of trees with plenty of undercover, thickets, and coppices. Pretty generally throughout England and Wales, but more sparingly distributed in Scotland and Ireland. The position of the nest and the skulking, mouse-like habits of its owner make it very difficult to find.

Materials.—Strong dry grass and moss, with an inner lining of finer grass. The nest is pretty deep and well built.

Eggs.—Four to seven, pale rosy-white, profusely

spotted and speckled all over with reddish-brown. Sometimes the markings are more numerous at the larger end, and occasionally a few thin, hair-like streaks are present. Size about ·72 by ·54 in.

Time.—May, June, and July.

Remarks.—Migratory, arriving in April and May, and departing in September. Notes: call, *tic, tic;* song, a chirping noise, similar to that made by a grasshopper, but louder and longer. Local and other names: Reeler, Cricket-bird, Grasshopper Lark. Leaves the nest quietly and quickly, and hides in the surrounding undergrowth.

—

WARBLER, MARSH.

Description of Parent Birds.—This bird very closely resembles the Reed Warbler, and it is only within recent years that it has been admitted to be a distinct British breeding species. Mr. Harting has done much to establish this fact, and specimens have been seen and procured in different parts of the country. Length about five and a half inches. Bill shorter and broader than in the case of the Reed Warbler, nearly straight, dark brown above, and pale brown below. Irides hazel. Mr Seebohm, who has had special facilities for examining specimens, describes the bird as follows, in his admirable work on British Birds :—

" The Marsh Warbler has the general colour of the upper-parts varying from olive-brown in spring plumage to earthy brown in summer, with a scarcely perceptible shade of rufous after the autumn moult, slightly paler on the rump. The eye stripe is nearly obsolete, and the innermost secondaries have broad, ill-defined pale edges. The breast, flanks, and under

tail-coverts are pale buff, shading into nearly white
on the chin, throat, and the centre of the belly.
. . . . Legs, feet, and claws horn colour."

Dresser and Sharpe say that this Warbler has
the legs of a pale flesh-brown, and that those of
the Reed Warbler are dark slaty-brown.

The female resembles the male, but is somewhat
smaller in size.

Situation and Locality.—A celebrated Continental
authority says that the nest is situated in low bushes,
overgrown with nettles, reeds, and other plants, and
that unlike the Reed Warbler, which builds its nest
amongst the reeds growing from the water, this bird
builds its nest amongst vegetation growing from the
bank of a stream or pond, and is never situated
over water. The nest is placed from a few inches to
several feet from the ground in swamps and other
places affording plenty of rough undergrowth cover.
It has been met with in the West of England and
in the Fen country.

Materials.—Dry grass-stems, dead leaves, moss,
and downy-fibre, with a lining of horsehair. The
nest is said not to be so deep as that of the Reed
Warbler, and to lack the wool which is so often
used by the last-named bird.

Eggs.—Four to seven, varying in ground colour
from greenish-white to greenish-blue, moderately
clouded and spotted with olive-brown, and underlying
markings of grey. The spots vary in size, intensity,
quantity, and disposition, but are generally most
numerous at the larger end of the egg. Their paler
ground colour generally distinguishes them from
those of the Reed Warbler. Size about ·72 by
·54 in.

Time.—June and July.

Remarks.—Migratory, arriving in May and departing in August. Notes: call and alarm, very similar to those of the Reed Warbler, but the song is said to be far finer, more melodious, and varied. It is delivered during the night in a similar way to that of the Nightingale. Local and other names, none. Not a very close sitter, but wonderfully adroit in slipping off the nest and hiding in surrounding vegetation.

WARBLER, REED. *Also* REED WREN.

Description of Parent Birds.—Length about five and a half inches. Bill fairly long, strong, nearly straight, dark horn colour on the upper mandible, and lighter on the under, which is yellowish at the base. Irides light yellowish-brown. A streak of cream colour runs from the base of the beak over the eyes. Head, neck, back, wings, rump, upper tail-coverts, and tail-quills pale brown, with a tinge of chestnut, which is most pronounced on the rump ; wing-quills dusky, and bordered with pale brown. Chin and throat white ; breast, flanks, and under tail-coverts white, tinged with cream colour ; belly white. Legs and toes dusky or slaty brown.

The female is rather smaller than the male, but similar in plumage.

Situation and Locality.—The nest is slung or suspended between the stems of reeds, at varying heights above the water. It is supported generally by three reeds, but upon occasion by two, four, or even five. Specimens may sometimes be met with amongst the branches of willow and other trees, growing near sluggish water. In reed beds, osier beds, and other places where suitable cover may be found, on the banks of ponds, reservoirs, and sluggish streams, principally on the eastern side of England.

It does not breed in the extreme western peninsula of England, and is rare in the northern counties. It is not known to breed in Scotland or in Ireland, and is said to be somewhat rare in Wales.

Materials.—Long blades of dried grass, seed-branches of reeds, roots, dry leaves, and wool, lined with fine grass and hairs. The nest is very deep, a necessity occasioned by its supports being swayed to and fro by gusts of wind.

Eggs.—Four to five, dull greenish-white, greyish-green, or pale greenish-blue, spotted, blotched, and blurred with darker greyish-green and light brown. A few black spots or streaks of dark brown are sometimes present. They are variable, both in the tint of the ground colour and markings. Size about ·74 by ·53 in. Their darker ground colour and the situation of the nest distinguish them from the eggs of the Marsh Warbler.

Time.—End of May, June, and even at the beginning of July eggs may be found.

Remarks.—Migratory, arriving in April and May and leaving in September. Notes : varied, loud, and hurriedly delivered. Naumann represents them as *tiri, tier, züch, zerr, scherk, heid, tret,* each note being repeated a number of times. Local and other names: Night Warbler, Reed Wren. Sits rather closely, and is noisy when disturbed.

— —

WARBLER, SAVI'S.

This bird used formerly to breed in the Fen country, but has long since ceased to do so on account of the drainage carried out therein, and is now only a very rare summer visitor.

WARBLER, SEDGE.

Description of Parent Birds.—Length four and three-quarter inches. Bill fairly long, straight, pointed, and brown, yellowish at the base of the under mandible. Irides brown. Crown of the head streaked with light and dark brown longitudinal lines; back of neck, back, and wing-coverts light reddish-brown, mixed with a darker tint of the same colour; wing-quills dark brown, bordered with lighter tinge; rump and upper tail-coverts tawny; tail-quills brown, indistinctly barred; from the base of the beak a yellowish-white streak runs over the eye and ear-coverts, the latter of which are brown. Chin and throat white; breast, belly, and under tail-coverts pale buff; under-side of tail-quills dusky brown; flanks bright buff. Legs, toes, and claws pale brown.

The female is darker on the under-parts, and less rufous on the under tail-coverts.

Situation and Locality.—Amongst thick, coarse, climbing herbage, brambles, wild rose, and other bushes, near streams, rivers, and swamps, pretty generally throughout the British Isles. Our illustration was procured near to Leatherhead, where the bird is numerous.

Materials.—Grass, coarse bents, and bits of moss, sometimes none of the latter whatever, lined internally with horsehairs, and occasionally with willow down. It is a deep, cup-shaped, loosely-built structure.

Eggs.—Five to six, pale yellowish or umber brown, sometimes a little clouded, suffused, or mottled with darker brown, and often streaked at the larger end with a few short, hairlike, black lines.

Variable, both in coloration and size. Measurements about ·67 by ·52 in.

Time.—May and June.

Remarks.—Migratory, arriving in April and May and departing in September. Notes: call, harsh and frequently uttered. The song of the male is loud, merry, imitative, and uttered often, during the night as well as the day. Local and other names: Chat, Sedge Bird, Sedge Wren, Reed Fauvette. Sits rather close, and slips quietly off.

WARBLER, WOOD. *Also* WOOD WREN.

Description of Parent Birds.—Length about five inches; bill rather short, slender, slightly curved, and brown; crown, nape, lesser wing-coverts, back, and upper tail-coverts olive-green, tinged with yellow; wings and tail dusky, bordered with yellow of varying shades. A line of bright primrose yellow runs from the base of the bill over the eye and ear-coverts; cheeks, chin, throat, and breast yellow; belly, vent, and under tail-coverts white; legs, toes, and claws brown. It is distinguished from the Willow Warbler by its broader yellow band over the eye, greener upper- and whiter under-parts, and longer wings.

The female is said to be a trifle larger than the male, but is similar in plumage.

Situation and Locality.—On the ground amongst thick herbage, in old plantations, woods, and other places well supplied with tall trees. Scattered generally throughout England, but said to be most numerous in some parts of Yorkshire and Durham. Met with in Scotland, but rare in Ireland.

Materials.—Dead grass, moss, and leaves, lined

SEDGE WARBLER.

with fine grass and horsehair. It is oval, and
domed like those of the Chiffchaff and Willow
Wren ; but is distinguished always from them by
having no feathers as an inner lining.

Eggs.—Five to seven, generally six; white in
ground colour, thickly spotted and speckled all
over with dark purplish-brown and ash-grey, most
thickly at the larger end. Size about ·65 by ·56 in.

Time.—May and June.

Remarks. — Migratory, arriving in April and
departing in October. Notes: song, *twee*, *twee*,
chea, *chea;* call note, *dee-ur.* Local and other
names : Wood Wren, Yellow Wren. Sits closely.

WARBLER, WILLOW. *Also* WILLOW WREN.

Description of Parent Birds.—Length about five
inches ; bill rather short, slender, slightly curved,
and brown in colour; under mandible paler at the
base. Irides hazel. Crown, neck, back, and upper
tail-coverts dullish olive-brown; wings and tail
dullish slate-brown, the feathers of the former being
bordered with olive-green. A pale yellow line runs
over the eye and ear-coverts; chin, throat, and
breast whitish yellow ; belly, flanks, and lower tail-
coverts greyish-white, slightly tinged with yellow;
legs, toes, and claws light brown.

The female is very similar in all respects to
the male. The bird is larger than the Chiffchaff,
and the feathers in its nest readily distinguish it
from that of the Wood Wren.

Situation and Locality.—On the ground amongst
coarse grass and weeds, entwining themselves round
slender twigs of low, open bushes growing on banks.
I have found it most frequently on banks near

WILLOW WARBLER.

W

willow and alder-fringed streams whilst trout-fishing,
and have often sat and watched the hen hop about
restlessly, and after a great deal of timid hesitation,
re-enter her nest. I was shown two nests in Westmoreland during June, 1894, in holes in walls.
One was at least three feet from the ground and
the other about a couple, not reckoning a high
bank upon which the wall stood. Throughout the
British Isles, wherever trees and bushes are to be
found in sufficient quantities. Our illustration is
from a photograph taken in a Yorkshire dale. The
front of the nest was opened so as to show the
eggs inside.

Materials.—Dead grass, moss, dead fern-fronds
and leaves, lined with horsehair, cowhair, and
liberal quantities of feathers. It is dome-shaped,
with a hole in front which is somewhat larger
than that of the Chiffchaff.

Eggs.—Four to eight, generally six to seven ;
white, spotted variably with pale rusty-red. Sometimes the spots are small and scattered pretty
evenly over the surface; at others they are larger,
less numerous, and more thickly congregated round
the larger end. Pure white and unspotted specimens
have been met with. The pale rusty-red markings
distinguish the eggs of this bird from those of the
Wood Wren and Chiffchaff. Size about ·64 by
·47 in.

Time.—April, May, June, and July, although
the last-named month is late. I once found a bird
sitting on eggs as late as the 4th of August.

Remarks.—Migratory, arriving in March and
April, and departing in August and September
according to some authorities, and October according to others. Specimens have been seen during
the winter in the southern counties of England.

Notes, long and shrill. Local and other names: Oven Bird, Jinney Wren, Scotch Wren, Yellow Wren, Hay Bird (a name also applied to the White-throat), Huckmuck (also applied to the Long Tailed Tit), Ground Wren. A close sitter.

WATERHEN. *See* MOORHEN.

WHEATEAR.

Description of Parent Birds.—Length about six inches. Bill fairly long, strong, and black, with a few bristles at the base. Irides hazel. Crown, nape, and back bluish-grey, tinged with light brown; rump and upper tail-coverts white. Wings nearly black, some of the feathers edged and tipped with buff. Tail-quills, upper two-thirds white, the remaining third black and broad. From the base of the beak, through the eye to the ear-coverts, is a band of black, over which is one of white, running from the forehead. Chin and throat dull white; breast pale cream colour, turning to a dull yellowish-white on the remainder of the under-parts. Legs, toes, and claws black.

The female is somewhat similar, except that she is browner on her upper-parts.

Situation and Locality.—Holes in dry walls, heaps of stones, old mine hillocks, under lumps of stone jutting from steep hillsides, in chinks of rock, quarries, peat stacks, and occasionally in rabbit-burrows. Our illustration is from a photograph of an old mountain limekiln in Westmoreland, which contained two nests, one inside and the other out, that is, one having its entrance from the inside and the other from the outside. On high moorland and uncultivated districts, bare of trees but abounding in rocks. To be met with in suitable districts over the whole of the British Isles, but most numerous in the North of England, Wales, Scotland, and Ireland.

Materials.—Roots, dead grass, moss, lined with wool, hair, rabbits' down, and feathers, loosely and clumsily put together. The materials named are, of course, not all present in the same nest, but are found according to the facilities the bird may enjoy for picking them up.

Eggs.—Four to seven, more generally five or six, of a pale greenish-blue, unspotted. I have found specimens nearly white sometimes, and they are said to be met with occasionally with a few small rusty spots on the larger end. Size about ·83 by ·61 in.

Time.—April, May, and June.

Remarks.—Migratory, arriving in March and departing in August or September, stray individuals sometimes lingering as late as December. Notes : *chick-chack-chack.* Local and other names : White Rump, Fallow Chat, Fallow Smick, Chacker, Chackbird, Clodhopper, Fallow Finch. A close sitter.

WHIMBREL.

Description of Parent Birds.—Length about sixteen inches. Bill long, slender, curved downward, and brown ; dark at the tip, and lighter towards the base. Crown dark brown, with a light central streak, and another passing from the base of the beak over the eye and ear-coverts ; neck brownish-grey, streaked with dark brown. Wings

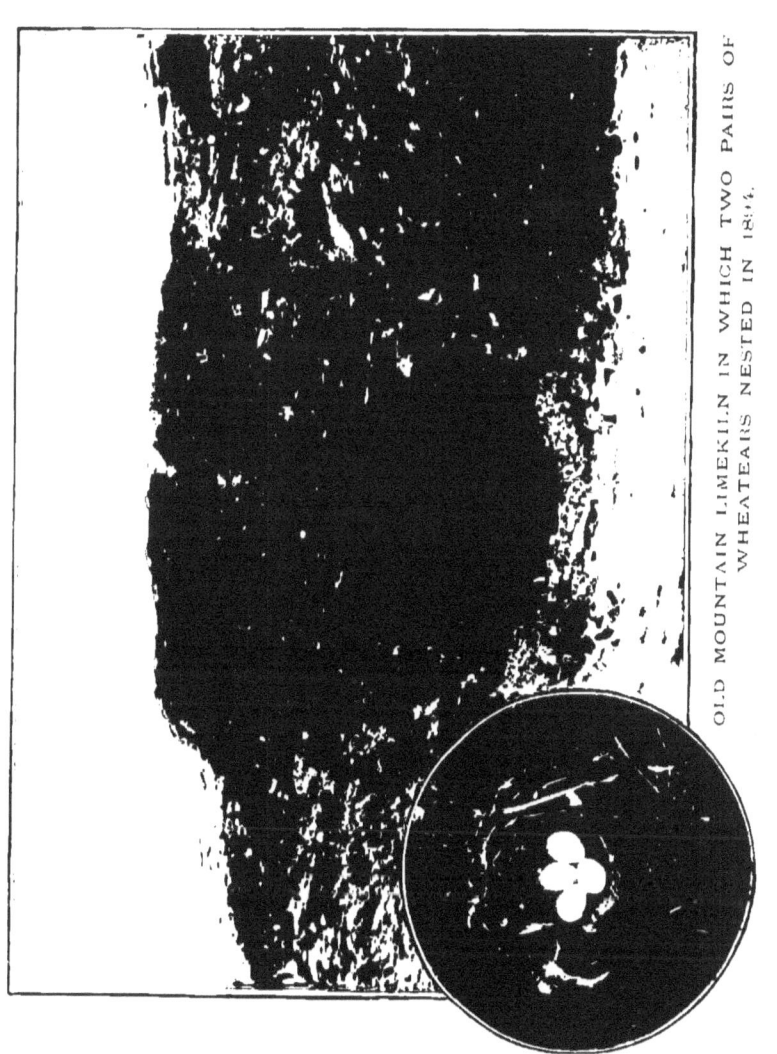

OLD MOUNTAIN LIMEKILN IN WHICH TWO PAIRS OF
WHEATEARS NESTED IN 1894.

dark brown, the feathers being margined and spotted
with pale brown and white; rump white, streaked
sparingly with brown. Tail-quills ash-brown, barred
with darkish brown. Chin white: upper breast
light brown, streaked with a darker tinge: lower
breast and belly almost white; flanks dull white,
barred with brown. Lower tail-coverts white,
streaked with brown. Legs and toes bluish-black:
claws black. Distinguished from the Curlew by
being considerably smaller in size.

The female is similar in plumage, but about two
inches greater in length.

Situation and Locality.—On the ground, amongst
the heather, or in the shelter of a tuft of grass,
on open moors in the Shetlands, Orkneys, and
Hebrides.

Materials.—A few blades of dried grass, used
as a lining to the depression chosen.

Eggs.—Four, pear-shaped, olive-green to olive-
brown in ground colour, blotched and spotted with
varying shades of brown and light grey. They
resemble the darker varieties of the Common
Curlew's, but are smaller, and are rather larger and
more pear-shaped than those of Richardson's Skua
with which they are likely to be confused. Size
about 2·35 by 1·65 in.

Time.—May and June.

Remarks.—Migratory, arriving at its breeding
grounds in April and May and departing in September. Notes: *tetty, tetty, tetty, tet.* Local and
other names: Whimbrel Curlew, Little Whaap,
Lang Whaap, Jack Curlew, Half Curlew, Curlew
Knot, Maybird, Titterel, Seven Whistler, Stone
Curlew (a name also given to the Norfolk Plover
or Thicknee), Chequer Bird. Not a close sitter,
but makes a great outcry when disturbed.

WHINCHAT.

Description of Parent Birds.—Length about five and a quarter inches; bill rather short, straight, and shiny black. Irides brown. Crown, nape, back, and smaller wing-coverts of two shades of brown, the feathers being dark in the centre and light round the edges; wings dark brown, the secondaries and tertials being edged with a paler hue; bastard or spurious wing white; upper half of tail white; lower dark brown, edged with a paler tinge of the same colour. From the base of the upper mandible, over the eye and ear-coverts, is a lengthy and rather broad streak of white. From the gape to the eye buff; chin white, extending beyond the lower margin of the ear-coverts; throat and breast light chestnut, turning to pale buff on the belly, vent, and under tail-coverts; legs, toes, and claws black.

In the female the white over the eye and on the wing is less distinctly marked, and the breast and belly are not so rich.

Situation and Locality.—On or near the ground in grass-fields, rough pasture land, on commons, at the foot of gorse bushes amongst the thick, tangling grass and dead lower branches; also amongst heather and coarse herbage. I know several small pastures in Yorkshire where pairs nest regularly year after year. The bird is extremely local, and very wary concerning the betrayal of its nest, which is well hidden; but I must confess I have not noticed the track or maze it has been reported to make as an approach to its nest. Our illustration is from a photograph taken behind Oban.

Materials.—Dead grass and moss, with an inner lining of horsehair.

Eggs.—Four to six, of a beautiful greenish-blue, sparingly speckled round the larger end with minute spots of reddish-brown. They are more attenuated, of a deeper blue, and less richly and clearly marked than the eggs of the Stonechat. The appearance of the parent birds, which are not often met with close together, will, however, readily settle any doubts. Size about ·76 by ·57 in.

Time.—May and June.

Remarks.—Migratory, arriving in April and departing in September or October. Note: *ü-tac.* Local and other names: Furze Chat, Grass Chat. Sits lightly, and, as mentioned above, is extremely wary. I detected the one above mentioned by watching through my field-glasses, the female go on to her nest, and then directing my brother to the spot by signs.

WHITETHROAT.

Description of Parent Birds.—Length about five and a half inches; bill somewhat short, straight, and brown, lighter towards the base of the under mandible. Irides yellowish. The whole of the upper-parts are brown, greyish on the head and neck, and reddish on the other parts: wing-quills greyish-brown, some of the smaller ones being edged with reddish-buff; tail-quills dull brown, some of the outer ones being edged and tipped with dirty white; chin and throat white; breast, belly, and under-parts generally, pale grey, tinted with a beautiful rosy flesh colour; legs, toes, and claws brown, lightest on the first named.

The female is less distinct in coloration, lacking

WHINCHAT.

the grey on the head and neck and the rosy tinge on her under-parts.

Situation and Locality. — Amongst nettles, brambles, thick rough grass, wild rose bushes, on hedgebanks in woods; on banks of streams, and wooded commons in nearly all suitable localities throughout the United Kingdom. One of those in our illustrated page was in a hedgebank, and the other, containing one egg and one newly-hatched young one, in a small open bush several feet away from a hedge. The nest was so much canted over on one side that the egg and young one were in danger of falling out.

Materials.—Dry grass stems and hair. The nest is deep, very flimsily constructed, and loosely attached.

Eggs. — Four to six, a dirty greenish-white, spotted and speckled with grey and brown. The spots are larger but not so pronounced as those of the Lesser Whitethroat; nor do they so often form a zone at the larger end, according to my experience. The ground colour is not of so clear a white either. Size about ·72 by ·55 in.

Time.—May, June, and July. The photograph of the nest with the young one in was taken in the last month.

Remarks. — Migratory, arriving in April and departing in September or October: individual specimens have, however, been observed as late as December. Notes: song "consists of numerous agreeable strains given in rapid succession" whilst the bird is in the air. Local and other names: Nettle Creeper. Sits closely.

WHITETHROATS' NESTS.

WHITETHROAT, LESSER.

Description of Parent Birds.—Length from five to five and a quarter inches; bill rather short, yellowish-brown at the base, and nearly black towards the tip. Irides from yellowish to pearl white, according to the age of the bird. Head, neck, back, rump, and upper tail-coverts greyish-brown; wing and tail-quills dusky, edged with greyish-brown; ear-coverts dark greyish-brown; chin, throat, and under-parts greyish-white, tinged between the thighs and round the vent with red, also, in some specimens, across the breast; legs, toes, and claws leaden grey.

The female is said to be a trifle duller in her plumage than the male, but in all other respects is very similar. Differs in size as its name implies, from the Whitethroat.

Situation and Locality.—In a low, sloping hedge (our illustration is from a photograph of a nest in such a situation and was taken in Surrey), amongst briars, brambles, nettles, gorse and low bushes, entangled by tall coarse grass and weeds; in gardens, orchards, on commons, rough waste lands, by river banks, and the sides of small woods. Fairly plentiful in the South and East of England, but rare in the West, North, and Scotland, and very rarely met with, indeed, in Ireland.

Materials. — Dead grass stalks, with an inner lining of horsehair. The whole structure is but a shallow, frail network-looking affair, that can be seen through with ease. It is sometimes tied or cemented together with old cobwebs.

Eggs.—Four to five ; white, light creamy white, or white with the faintest suggestion of green, in

LESSER WHITETHROAT.

ground colour, spotted and speckled with ash-grey, greenish-brown, and umber-brown. The markings generally form a belt round the larger end. Size about ·66 by ·52 in. Distinguished by small size, clean ground colour, and bold belt-inclining spots

Time.—May, June, and July.

Remarks. — Migratory, arriving in April and leaving in September. Notes: call, *check*, repeated several times, and an incessant chattering, sometimes loud and grating, at others low and not unpleasant. Local or other names: none. Sits closely.

WIGEON.

Description of Parent Birds.—Length about eighteen inches; bill rather short, narrow, highish at the base, and bluish-grey, tipped with black. Irides dark brown. Forehead and crown cream colour; rest of head and upper part of neck rich bay, almost black on the chin and throat. A streak of green passes backward from the eye, and the cheeks and neck are minutely spotted with blackish-green; back and scapulars greyish-white, barred with fine irregular lines of black; wing-coverts white, tipped with black; outer webs of secondaries green, tipped with black; tertials or inner secondaries black on the outer webs, which are broadly margined with white; primaries dusky brown; upper tail-coverts freckled with grey; tail feathers long, wedge-shaped, and dusky-black, mixed with grey and brown; lower part of the neck and shoulders pale red; breast, belly, and vent white; sides and flanks marked with fine undulating lines of black; under tail-coverts rich black; legs, toes, and webs dark brown.

The female is a little smaller; her head and neck are brown, tinged with red and speckled with dark brown. The feathers on her upper-parts are dark brown, margined with light reddish-brown; breast pale brown; under-parts almost white; wings and tail somewhat similar to those of the male.

At the end of the spring and beginning of summer the male retires to some solitary swamp, and casts off his gay dress and assumes a dull and sombre one, which he continues to wear until the autumn.

Situation and Locality.—On the ground, in a clump of rushes, tuft of heather, or amongst coarse grass, flags, reeds, and under dwarf bushes, cleverly concealed, in the neighbourhood of lochs. swamps, tarns, or rivers, where the ground is rough and shelter-affording; in suitable parts of Scotland, the Orkneys and Shetlands, and probably in one or two parts of Ireland.

Materials.—Reeds, decayed rushes, leaves, and dry grass, with an inner lining of down from the bird's own body. The tufts are dark sooty-brown, with conspicuous white tips.

Eggs.—Six to twelve, generally seven or eight; creamy white, somewhat like those of the Gadwall. The locality of the nest, and the feathers of the down tufts prevent any chance of confusion with that duck. Size about 2·2 by 1·5 in.

Time.—May.

Remarks.—A winter visitor, generally arriving in September or October, and departing North in March or April. Notes, a shrill whistle. Local and other names: Whew Duck, Whewer, Pandle Whew, Yellow Poll, Easterlings, Whim. A close sitter.

WOODCOCK.

Description of Parent Birds.—Length about fifteen inches. Bill long, straight, dark brown at the tip, and pale reddish-brown at the base. Irides dark brown. A dark streak of brown extends from the gape to the eye. Head and upper-parts a mixture of rusty brown, black, and grey, which occur in each feather and produce a handsomely variegated effect. Cheeks and the whole of the under-parts yellowish-white, numerously barred with dark wavy lines. Legs and toes brown; claws black.

The female is similar in plumage to the male, but both are subject to great variation in size and colour.

Situation and Locality.—On the ground, amongst dead grass, under brackens, ferns, brambles, and sometimes amongst dead leaves at the foot of a tree; in woods, plantations, coppices, and forests with plenty of the undergrowth just named; very sparingly but pretty generally in suitable localities throughout the United Kingdom. Our illustration is from a photograph of a nest situated under a fallen branch in a Norfolk wood.

Materials.—Dry grass, fern-fronds, and dead leaves, placed in some natural, dry, and sheltered hollow; sometimes a suitable declivity is scraped by the bird.

Eggs.—Four, yellowish-white to buffish-brown. blotched with pale chestnut-brown and ash-grey. Size about 1·7 by 1·35 in.

Time.—March, April, and May.

Remarks.—A winter visitor, although a few individuals stay with us all the year round. Notes: alarm, *skaych*, somewhat resembling that of the

WOODCOCK.

X

Snipe. Local and other names, none. Harmonises
in appearance with its surroundings, and sits closely.

WOOD LARK.

Description of Parent Birds.—Length about six
inches. Bill of medium length, straight, and dusky
brown, lighter at the base of the under mandible.
Irides hazel. Crown light brown, streaked with a
darker shade of the same colour; feathers form an
erectile crest. Over the eye and ear-coverts is a
streak of pale yellowish-brown. The upper parts of
the body are of a light reddish-brown, streaked
and patched with dusky on the neck and back.
Wing-quills dusky, bordered with brown. Tail-
coverts long and brown. Tail short, outer feathers
on either side brownish-black, tipped with dirty
white. Throat, breast, belly, and vent yellowish-
white, tinged with brown, the first-named being
sparingly speckled and streaked with a darker hue;
breast streaked and spotted more thickly with the
same colour. Legs, toes, and claws brown; hind
claw long and curved.

The female is smaller than the male, and is
said to be yellower on the breast and to have
larger markings. Distinguished from the Skylark
by its more slender bill and shorter tail.

Situation and Locality.—On the ground, usually
well concealed by a tuft of grass, low plant, or
bramble; sometimes at the foot of a tree or on the
side of a bank, in fields and pastures, on commons
and heaths adjoining woods, copses, and plantations,
most numerously in the southern counties, occasionally in the north, and rarely in Scotland and
Ireland.

Materials.—Coarse grass on the outside, finer

grass, moss, and hair as an inner lining. The nest is placed in a little hollow, either natural or scratched out by the bird.

Eggs.—Four to five, pale greenish-white, light brownish-yellow, or pale reddish-white in ground colour, thickly speckled and spotted with dull reddish-brown, and underlying markings of dark grey. The markings sometimes form a zone at the larger end. Size about ·84 by ·65 in. Distinguishable from those of the Skylark by small reddish-brown spots and lighter and less obscured ground colour.

Time.—March, April, May, and June.

Remarks. — Resident and migratory. Notes : sings on the wing and perched on the boughs of trees ; call, uttered constantly during flight, *tweedle, weedle, weedle.* Local and other names : none. A close sitter.

WOODPECKER, GREATER SPOTTED.

Description of Parent Birds.—Length nearly nine and a half inches. Beak of medium length, straight, sharp at the tip, and dusky. Irides red. Forehead buffish ; round the eyes and ear-coverts dirty white. Crown black ; back of head bright scarlet. A black stripe commences at the gape and, widening, passes backward under the eye and ear-coverts to the nape ; another commences on the side of the throat and, passing backwards, also meets the black on the back of the neck. A horizontal, elongated patch of white is enclosed by the black on either side of the neck ; the back, rump, and upper tail-coverts are black. Wings black, variegated with white spots and a large patch of the same colour on the scapulars.

Tail longest in the centre, and black, the outside feathers being tipped in increasing lengths from the middle with white. Chin, throat, breast, and belly dirty white; vent and under tail-coverts bright scarlet. Legs, toes, and claws greenish-grey.

The female is a little smaller than the male, and lacks the scarlet on the back of her head. Easily distinguished from the Lesser Spotted Woodpecker by its much greater size and lack of white on the back.

Situation and Locality.—In holes in trees, either dug by the bird's own exertions, or a decayed hole in the trunk or a branch, adapted and enlarged. It is somewhat similar to that of the other Wood-peckers, and varies from ten to twenty inches in depth. In forests, well-timbered parks, woods, and other places where old trees exist. It is found in nearly all the counties of England and Wales, excepting those north of Yorkshire, where it is an exceedingly rare bird, as it also is both in Scotland and Ireland. Our illustration is from a photograph procured in Middlesex.

Materials.—None, the eggs being laid on the powdered wood and chips produced in making the cavity for their reception.

Eggs.—Four to seven, occasionally as many as eight, white, unspotted, and glossy. Size about 1·05 by ·75 in. Distinguished by their size and characteristics of parent birds.

Time.—May and June.

Remarks.—Resident, but its numbers are said to be increased in winter by Continental visitors. Notes: *gich-gich, quet-quet, tra, tra, tra.* Local and other names: Witwall, Woodnacker, Wood-pie, French-pie, Great Black and White Woodpecker, Spickel-pied Woodpecker. A very close sitter.

SITE OF A GREATER SPOTTED WOODPECKER'S NEST.

WOODPECKER, GREEN.

Description of Parent Birds.—Length about
thirteen inches. Beak rather long, strong, and
dusky in colour. Irides greyish-white. Crown,
crimson ; neck, back, lesser wing-coverts, and
scapulars green; rump yellow ; upper tail-coverts
yellow, tinged with green in parts. Wing-quills
dusky, barred and spotted with buffish-white, some
of the lesser being margined with olive-green.
Tail-quills dusky, barred with greyish-brown, some
of them being margined with green. From the
base of the beak, round and behind the eyes, the
feathers are black. A crimson streak, bordered
with black, runs from the gape some little way
down the sides of the neck. Chin, throat, breast,
and all under-parts pale greyish-green. Legs, toes,
and claws ash-colour. The toes are disposed, two
in front and two behind, and claws hooked.

The female has less crimson on her crown, and
none at all from the gape down the sides of the
neck, which is black.

Situation and Locality.—In holes in trees,
generally dug by the bird's own exertions, those
composed of soft wood being preferred. The hole
is from ten to eighteen inches deep. It breeds in
suitably wooded localities nearly all over England,
but is least numerous in the northern counties, and
does not breed in either Scotland or Ireland.

Materials.—Only the chips and bits of decayed
wood, detached in hewing the nesting-place.

Eggs.—Five to seven, sometimes eight, pure
white, unspotted, and glossy. Size about 1·3 by
·92 in. Distinguished by large size and appearance
of parent birds.

Time.—April and May.

Remarks.—Resident. Notes, several, which have been represented as follows:—male spring note, *tiacacan, tiacacan* ; call, used all the year round, *pleu, pleu, pleu.* Some writers represent the call as *yaffa, yaffa, yaffle.*' Local and other names: Rainbird, Popinjay, Awlbird, Yaffle, Tongue Bird, Gally, Rain-fowl, Pick-a-tree, Whetile, Woodwale, Wood-speight, Yaffingale. A close sitter, often occupying the same nest year after year, when not stolen by Starlings.

WOODPECKER, LESSER SPOTTED.

Description of Parent Birds.—Length about five and three-quarter inches. Beak of medium length, broad at the base, straight, and leaden grey in colour. Irides hazel. Crown bright scarlet, sides of head brownish-white. A black stripe runs from the base of the beak over the eye to the nape, which is black also ; another runs from the base of the under mandible below the eye, and beneath the ear-coverts. Back of neck and upper back black. Wings black, barred and spotted with white ; middle of back white, barred with black ; rump and upper tail-coverts black. Tail-quills black, some of them edged and tipped with white, others white, barred with black. Chin, throat, and under-parts brownish-white ; sides of breast and flanks streaked and slightly barred with black. Legs, toes, and claws lead grey.

The female has the crown brownish-white, occasionally shaded with red ; black of nape commences further forward, and under-parts are darker.

Situation and Locality.—In a hole, dug or enlarged, in the stem or large branch of a tree. Pear and apple trees appear to be favourites. The hole is usually from six or seven to twelve or fourteen inches deep. In orchards, spinneys, parks, woods, and well-timbered districts generally, in the South and West of England, and as far north as York.

Materials.—None, the eggs being laid on the wood dust and fine chips at the bottom of the hole.

Eggs.—Five to nine, six being the general number, white and glossy. Size about ·76 by ·58 in. Distinguished from those of the Wryneck only by smaller size and a sight of parents.

Time.—April and May.

Remarks.—Resident. Notes: *tic-tic*, or *kink-kink*. Local and other names: Crank Bird, Pump-borer, Least-spotted Woodpecker, Little Black and White Woodpecker, Hickwall, Barred Woodpecker. Sits closely.

WREN, COMMON.

Description of Parent Birds.—Length a little under four inches; bill of moderate length, slightly curved downwards, dark brown on the top, and light brown underneath. Irides hazel. A dull, whitish line runs over the eye and ear-coverts. Head, nape, back, and rump reddish-brown, faintly marked on the two latter with wavy bars of light and dark brown; wing and tail feathers marked with light reddish-brown and black bars; chin, cheeks, throat, and breast greyish-white, tinged with buff; belly, sides, and thighs light brown, marked with narrow wavy bars of a darker tint of the same colour; under tail-coverts indistinctly spotted with black and dirty white; legs, toes, and claws pale brown.

The female is rather smaller, duller on her upper-parts, and darker beneath.

NESTS OF COMMON WREN.

Situation and Locality.—I have found the nest
of this bird in crevices and holes of rock, in the
gnarled roots of trees growing on the banks of
streams, amongst fern roots growing on the sides
of banks, in banks of rivers, amongst ivy growing
against walls and trees, in holes in brick and stone
walls, faggot and hay-stacks ; in the thatch of
barns ; amongst wreckage lodged in a thorn bush
by a high flood ; between the stems of two trees
growing close together over a stream, and lodged
amongst a few slender twigs sprouting from the
trunk of a tree where a large branch had been
lopped off, and various other situations. The subject of one of our illustrations was found near the
top of a hedge, and the other amongst ivy growing
against a garden wall. The nest has been found
in all sorts of odd places, such as in cabbages
run to seed, bodies of scarecrows, and the skeleton
of a Carrion Crow hanging against a wall. Generally
throughout the British Isles.

Materials.—Moss, dead leaves, fern-fronds, roots,
dry grass-stalks, the stems of leaves, lined with
hair and feathers. I remember once finding one
lined entirely with the feathers of a hen grouse,
whose skeleton I discovered not far away. The
bird has a very shrewd idea about the value of
harmonisation and mimicry. The nest I mentioned
finding amongst fern roots was composed outside
of dead fern-fronds, and the one between the stems
of two trees growing close together over a brook,
of bright green moss, matching that upon the trunk
on either side exactly. The one in the twigs
growing from the stem of a tree was composed
outside entirely of dead leaves, and looked exactly
like an accidental collection. The bird practises
the curious habit of building the outer structure

of several nests, but whenever one is found with an inner lining of feathers it is sure to be laid in.

Eggs.—Four to eight, generally from five to seven, although as many as twelve and fourteen, and even twenty, have upon rare occasions been found. White, sparingly spotted with brownish-red of varying shades, generally distributed at the larger end. Specimens are sometimes found quite unspotted. When I discovered the ivy-surrounded nest (figuring in the circular illustration) it contained three unspotted eggs, and I remarked to my brother that the layer was in a poor state of health. We photographed it on the Saturday afternoon, and as I was anxious to secure one of the eggs, I visited the nest at six o'clock on the following Monday morning, and found the hen inside. I went away for an hour, and when I returned she was still on her eggs, and I discovered that she was quite dead and in a very emaciated condition. On dissection there was no sign of further eggs in her body. Size about ·7 by ·51 in.

Time.—April, May, June, and July.

Remarks.—Resident. Notes: alarm, a jarring kind of note; song, loud, joyous, and heard all the year round. Local and other names: Cutty Wren, Titty Wren, Jenny Wren (a name also applied to the Chiffchaff and Willow Wren), Tom Tit (in the North of England), Kitty Wren. A close sitter.

WREN, ST. KILDA.

To Mr. Dixon, who gave it the distinguishing name here used, belongs the honour of first discovering the differences between this bird and the Common Wren. According to his observations—

and he is a very capable and reliable authority—
it is paler in colour, has longer feet, utters louder
and harsher notes, lays on an average bigger eggs,
and is not found anywhere outside the St. Kilda
group of islands. In other respects it is similar
in its appearance and habits to its near ally, the
Common Wren.

WREN, GOLDEN-CRESTED. *See* GOLDCREST.

WREN, REED. *See* WARBLER, REED.

WREN, WILLOW. *See* WARBLER, WILLOW.

WREN, WOOD. *See* WARBLER, WOOD.

WRYNECK.

Description of Parent Birds.—Length about
seven inches. Beak rather short, straight-pointed,
and brown. The whole of the upper parts of the
body consist of varying shades of brown, mixed
with grey, pencilled, mottled, barred, and streaked
with buff, greyish-white, brownish-black, and black.
The top of the head is barred with blackish-brown,
the nape of the neck striped with the same, and
also parts of the back and wings. The wing-quills
are dark brown, barred and spotted with two shades
of buff. Tail-quills greyish-brown, marked with
several irregular blackish-brown bars. All the
under-parts are dull white, 'tinged with yellowish-

buff on the chin, throat, flanks, and under tail-coverts, which are barred with dark brown. The breast and belly are marked with small triangular spots of dark brown. Legs, toes, and claws brown.

The female is somewhat duller in her coloration.

Situation and Locality.—In holes in trees, at varying heights from the ground and at differing depths. The deserted hole of a Woodpecker is a favourite site. In open woodlands, parks, trees growing by brooks, roads, and in fields; in the South and East of England. It is scarce in the west and north, and more so in Scotland and Ireland.

Materials.—Generally the decayed and powdered wood at the bottom of the hole selected. Occasionally it is said to contain other materials, such as moss, wool, hair, or feathers, but these might have been previously deposited by some other bird, as the Wryneck is reported to be not averse to using a hole so furnished.

Eggs.—Six to ten, generally seven or eight, pure white, unspotted, and often mistaken for those of the Lesser Woodpecker, from which they differ, however, in being a trifle larger. A sight of the parent birds is the only certain method of identification. Average size about ·85 by ·63 in.

Time.—May and June.

Remarks.—Migratory, arriving in April and leaving in September. Note: a betraying *peel, peel, peel,* uttered about nine times in unbroken succession. Local and other names: Snake Bird, Cuckoo's Mate, Tongue Bird, Emmet Hunter, Long Tongue. Sits close, and hisses.

YELLOW-HAMMER. *Also* YELLOW BUNTING.

Description of Parent Birds.—Length about seven inches. Bill short, strong, and bluish horn colour, tinged with brown. Irides dark brown. There are a few short bristles round the base of the bill. Head and nape yellow, tinged with green, and marked on the crown with a few streaks of dusky black and olive. Back bright reddish-brown, tinged with yellowish-green. Wing-quills dusky, bordered with greenish-yellow; rump bright chestnut; tail slightly forked, dusky black, edged with greenish-yellow, the two outer feathers being marked with white spots. Throat, breast, belly, and under tail-coverts bright yellow, the breast being sometimes marked with reddish spots, and the sides streaked. Legs, toes, and claws yellowish-brown.

The female is a trifle smaller, and much duller in her plumage. She is less yellow, and more thickly marked with brown. Her tail is also lighter, and has less white on the outsides. I have often been struck by her close harmonisation with surrounding objects when her nest is on the ground. Both sexes are subject to variation.

Situation and Locality.—On or near the ground, although specimens may sometimes be found at a height of eight or ten feet. In hedgebanks, amongst brambles, nettles, and coarse grass at the foot of light open bushes. Our illustration is from a photograph taken in Norfolk. On pieces of waste land, commons, pastures, grass-fields, and arable lands, in all suitable localities throughout the United Kingdom.

Materials.—Dry grass, roots, and moss, with an

YELLOW-HAMMER.

inner lining of fine grass and horsehair. The nest varies in bulk according to situation.

Eggs.—Three to six, generally four or five. Ground colour dingy white, tinged with purple, and streaked, veined, spotted, and blotched with dark purplish-brown, the streaks and lines generally terminating in a spot of the same colour. There are also underlying markings of purplish-grey. Subject to great variation. Size about ·88 by ·65 in. The purple tinge of the ground colour and the thick scribbling lines distinguish them from those of the Cirl Bunting, with which they are likely to be confused.

Time.—April, May, June, July, and August.

Remarks.—Resident. Notes: *chit, chit,* followed by a long, harsh *chire-r-r.* Bechstein represents the song by *te, te, te, te, te, te, tirgee,* but it is popularly interpreted in this country as *Bit o' bread and no chee-e-e-se.* Local and other names: Yellow Bunting, Yellow Yowley, Goldspink, Yoist, Yellow Yite, Yellow Yoldring, Yeldrock, Yellow Yeldring. A very close sitter.

NEWLY-HATCHED GULL.

PRINTED BY CASSELL & COMPANY, LIMITED, LA BELLE SAUVAGE, LONDON, E.C.

5·798

A SELECTED LIST

OF

CASSELL & COMPANY'S

PUBLICATIONS.

1 G 6.96

Illustrated, Fine Art, and other Volumes.

Adventure, The World of. Fully Illustrated. Complete in Three Vols. 9s. each.
Adventures in Criticism. By A. T. QUILLER-COUCH. 6s.
Africa and its Explorers, The Story of. By Dr. ROBERT BROWN, F.R.G.S., &c. With about 800 Original Illustrations. *Cheap Edition.* In 4 Vols. 4s. each.
Animal Painting in Water Colours. With Coloured Plates. 5s.
Animals, Popular History of. By HENRY SCHERREN, F.Z.S. With 13 Coloured Plates and other Illustrations. 7s. 6d.
Architectural Drawing. By R. PHENÉ SPIERS. Illustrated. 10s. 6d.
Art, The Magazine of. *Yearly Volume,* 21s. *The Two Half-Yearly Volumes for 1897 can also be had,* 10s. 6d. *each.*
Artistic Anatomy. By Prof. M. DUVAL. *Cheap Edition,* 3s. 6d.
Ballads and Songs. By WILLIAM MAKEPEACE THACKERAY. With Original Illustrations by H. M. BROCK. 6s.
Barber, Charles Burton, The Works of. With Forty-one Plates and Portraits, and Introduction by HARRY FURNISS. *Cheap Edition,* 7s. 6d.
Battles of the Nineteenth Century. An entirely New and Original Work, with Several Hundred Illustrations. Complete in Two Vols., 9s. each.
"Belle Sauvage" Library, The. Cloth, 2s. (*A complete list of the volumes post free on application.*)
Beetles, Butterflies, Moths, and other Insects. By A. W. KAPPEL, F.L.S., F.E.S., and W. EGMONT KIRBY. With 12 Coloured Plates. 3s. 6d.
Biographical Dictionary, Cassell's New. Containing Memoirs of the Most Eminent Men and Women of all Ages and Countries. *Cheap Edition.* 3s. 6d.
Birds' Nests, British: How, Where, and When to Find and Identify Them. By R. KEARTON, F.Z.S. With nearly 130 Illustrations of Nests, Eggs, Young, &c., from Photographs by C. KEARTON. 21s.
Birds' Nests, Eggs, and Egg-Collecting. By R. KEARTON, F.Z.S. Illustrated with 22 Coloured Plates of Eggs. *Enlarged Edition.* 5s.
Black Watch, The. The Record of an Historic Regiment. By ARCHIBALD FORBES, LL.D. 6s.
Britain's Roll of Glory; or, the Victoria Cross, its Heroes, and their Valour. By D. H. PARRY. Illustrated. 7s. 6d.
British Ballads. With 300 Original Illustrations. *Cheap Edition.* Two Volumes in One. Cloth, 7s. 6d.
British Battles on Land and Sea. By JAMES GRANT. With about 800 Illustrations. *Cheap Edition.* Four Vols., 3s. 6d. each.
Building World. In Half-Yearly Volumes, 4s. each.
Butterflies and Moths, European. By W. F. KIRBY. With 61 Coloured Plates, 35s.
Canaries and Cage-Birds, The Illustrated Book of. By W. A. BLAKSTON, W. SWAYSLAND, and A. F. WIENER. With 56 Facsimile Coloured Plates. 35s.
Cassell's Magazine. Half-Yearly Volumes, 5s. each; or Yearly Volumes, 8s. each.
Cathedrals, Abbeys, and Churches of England and Wales. Descriptive, Historical, Pictorial. *Popular Edition.* Two Vols. 25s.
Cats and Kittens. By HENRIETTE RONNER. With Portrait and 13 magnificent Full-page Photogravure Plates and numerous Illustrations. 4to, £2 10s.
China Painting. By FLORENCE LEWIS. With Sixteen Coloured Plates, &c. 5s.
Choice Dishes at Small Cost. By A. G. PAYNE. *Cheap Edition,* 1s.
Chums. The Illustrated Paper for Boys. Yearly Volume, 8s.
Cities of the World. Four Vols. Illustrated. 7s. 6d. each.
Civil Service, Guide to Employment in the. *Entirely New Edition.* Paper, 1s.; cloth, 1s. 6d.
Clinical Manuals for Practitioners and Students of Medicine. (*A List of Volumes forwarded post free on application to the Publishers.*)

Cobden Club, Works published for the. (*A Complete List on application.*)
Colour. By Prof. A. H. CHURCH. *New and Enlarged Edition*, 3s. 6d.
Combe, George, The Select Works of. Issued by Authority of the Combe Trustees. *Popular Edition*, 1s. each, net.
    The Constitution of Man. Moral Philosophy. Science and Religion. Discussions on Education. American Notes.
Conning Tower, In a; or, How I Took H.M.S. "Majestic" into Action. By H. O. ARNOLD-FORSTER, M.P. *Cheap Edition.* Illustrated. 6d.
Conquests of the Cross. Edited by EDWIN HODDER. With numerous Original Illustrations. Complete in Three Vols. 9s. each.
Cookery, Cassell's Dictionary of. With about 9,000 Recipes. 5s.
Cookery, A Year's. By PHYLLIS BROWNE. *New and Enlarged Edition*, 3s. 6d.
Cookery Book, Cassell's New Universal. By LIZZIE HERITAGE. With 12 Coloured Plates and other Illustrations. 1,344 pages, strongly bound in leather gilt, 6s.
Cookery, Cassell's Popular. With Four Coloured Plates. Cloth gilt, 2s.
Cookery, Cassell's Shilling. 135*th Thousand.* 1s.
Cookery, Vegetarian. By A. G. PAYNE. 1s. 6d.
Cooking by Gas, The Art of. By MARIE J. SUGG. Illustrated. Cloth, 2s.
Cottage Gardening. Edited by W. ROBINSON, F.L.S. Illustrated. Half-yearly Vols., 2s. 6d. each.
Countries of the World, The. By Dr. ROBERT BROWN, M.A., F.L.S. With about 750 Illustrations. *Cheap Edition.* Vols. I. to V., 6s. each.
Cyclopædia, Cassell's Concise. With about 600 Illustrations. 5s.
Cyclopædia, Cassell's Miniature. Containing 30,000 Subjects. Cloth, 2s.6d.; half-roxburgh, 4s.
Dictionaries. (For description, see alphabetical letter.) Religion, Biographical, Encyclopædic, Concise Cyclopædia, Miniature Cyclopædia, Mechanical, English, English History, Phrase and Fable, Cookery, Domestic. (French, German, and Latin, see with *Educational Works*.)
Diet and Cookery for Common Ailments. By a Fellow of the Royal College of Physicians and PHYLLIS BROWNE. *Cheap Edition.* 2s. 6d.
Dog, Illustrated Book of the. By VERO SHAW, B.A. With 28 Coloured Plates. Cloth bevelled, 35s.; half-morocco, 45s.
Domestic Dictionary, The. An Encyclopædia for the Household. Cloth, 7s. 6d.
Doré Don Quixote, The. With about 400 Illustrations by GUSTAVE DORÉ. *Cheap Edition.* Cloth, 10s. 6d.
Doré Gallery, The. With 250 Illustrations by GUSTAVE DORÉ. 4to, 42s.
Doré's Dante's Inferno. Illustrated by GUSTAVE DORÉ. *Popular Edition.* With Preface by A. J. BUTLER. Cloth gilt or buckram, 7s. 6d.
Doré's Dante's Purgatory and Paradise. Illustrated by GUSTAVE DORÉ. *Cheap Edition.* 7s. 6d.
Doré's Milton's Paradise Lost. Illustrated by GUSTAVE DORÉ. 4to, 21s. *Popular Edition.* Cloth gilt, or buckram gilt, 7s. 6d.
Earth, Our, and its Story. Edited by Dr. ROBERT BROWN, F.L.S. With 36 Coloured Plates and nearly 800 Wood Engravings. In Three Vols. 9s. each.
Edinburgh, Old and New, Cassell's. With 600 Illustrations. Three Vols. 9s. each; library binding, £1 10s. the set.
Egypt: Descriptive, Historical, and Picturesque. By Prof. G. EBERS. Translated by CLARA BELL, with Notes by SAMUEL BIRCH, LL.D.,&c. Two Vols. 42s.
Electric Current, The. How Produced and How Used. By R. MULLINEUX WALMSLEY, D.Sc., &c. Illustrated. 10s. 6d.
Electricity, Practical. By Prof. W. E. AYRTON, F.R.S. *Entirely New and Enlarged Edition.* Completely re-written. Illustrated. 9s.
Electricity in the Service of Man. A Popular and Practical Treatise. With upwards of 950 Illustrations. *New and Cheaper Edition.* 7s. 6d.
Employment for Boys on Leaving School, Guide to. By W. S. BEARD, F.R.G.S. 1s. 6d.
Encyclopædic Dictionary, The. Complete in Fourteen Divisional Vols., 10s. 6d. each; or Seven Vols., half-morocco, 21s. each; half-russia, 25s. each.
England and Wales, Pictorial. With upwards of 320 beautiful illustrations prepared from copyright photographs. 9s. Also an edition on superior paper, bound in half-persian, marble sides, gilt edges and in box, 15s. net.
England, A History of. From the Landing of Julius Cæsar to the Present Day. By H. O. ARNOLD-FORSTER, M.P. Fully Illustrated, 5s.

England, Cassell's Illustrated History of. From the earliest period to the present time. With upwards of 2,000 Illustrations. New Serial issue in Parts, 6d. each.

English Dictionary, Cassell's. Containing Definitions of upwards of 100,000 Words and Phrases. *Cheap Edition*, 3s. 6d. ; *Superior Edition*, 5s.

English History, The Dictionary of. Edited by SIDNEY LOW, B.A., and Prof. F. S. PULLING, M.A., with Contributions by Eminent Writers. *New Edition.* 7s. 6d.

English Literature, Library of. By Prof. H. MORLEY. In 5 Vols. 7s. 6d. each.

English Literature, Morley's First Sketch of. *Revised Edition.* 7s. 6d.

English Literature, The Story of. By ANNA BUCKLAND. 3s. 6d.

English Writers from the Earliest Period to Shakespeare. By HENRY MORLEY. Eleven Vols. 5s. each.

Æsop's Fables. Illustrated by ERNEST GRISET. *Cheap Edition.* Cloth, 3s. 6d. ; bevelled boards, gilt edges, 5s.

Etiquette of Good Society. *New Edition.* Edited and Revised by LADY COLIN CAMPBELL. 1s. ; cloth, 1s. 6d.

Fairy Tales Far and Near. Retold by Q. Illustrated. 3s. 6d.

Fairway Island. By HORACE HUTCHINSON. *Cheap Edition.* 2s. 6d.

Family Doctor, Cassell's. By A MEDICAL MAN. Illustrated, 10s. 6d.

Family Lawyer, Cassell's. An Entirely New and Original Work. By a Barrister-at-Law. 10s 6d.

Fiction, Cassell's Popular Library of. 3s. 6d. each.

Loveday. By A. E. WICKHAM.
Tiny Luttrell. By E. W. Hornung.
The White Shield. By Bertram Mitford.
Tuxter's Little Maid. By G. B. Burgin.
The Hispaniola Plate. (1693—1893.) By John Bloundelle Burton.
Highway of Sorrow. By Hesba Stretton and * * * * * * *, a Famous Russian Exile.
King Solomon's Mines. By H. Rider Haggard. (Also People's Edition, 6d.)
The Lights of Sydney. By Lilian Turner.
The Admirable Lady Biddy Fane. By Frank Barrett.
Out of the Jaws of Death. By Frank Barrett.
List, ye Landsmen ! A Romance of Incident. By W. Clark Russell.
Ia: A Love Story. By Q.
The Red Terror : A Story of the Paris Commune. By Edward King.
The Little Squire. By Mrs. Henry de la Pasture
Zero, the Slaver. A Romance of Equatorial Africa. By Lawrence Fletcher.
Into the Unknown ! A Romance of South Africa. By Lawrence Fletcher.
Mount Desolation. An Australian Romance. By W. Carlton Dawe.
Pomona's Travels. By Frank R. Stockton.

The Reputation of George Saxon. By Morley Roberts.
A Prison Princess. By Major Arthur Griffiths.
Queen's Scarlet, The. By George Manville Fenn.
Capture of the "Estrella," The. A Tale of the Slave Trade. By Commander Claud Harding, R.N.
The Awkward Squads. And other Ulster Stories. By Shan F. Bullock.
A King's Hussar. By Herbert Compton.
A Free-Lance in a Far Land. By Herbert Compton.
Playthings and Parodies. Short Stories. Sketches, &c. By Barry Pain.
Fourteen to One, &c. By Elizabeth Stuart Phelps.
The Medicine Lady. By L. T. Meade.
Father Stafford. By Anthony Hope.
" La Bella," and others. By Egerton Castle.
The Avenger of Blood. By J. Maclaren Cobban.
The Man in Black. By Stanley Weyman.
The Doings of Raffles Haw. By A. Conan Doyle.

Field Naturalist's Handbook, The. By Revs. J. G. WOOD and THEODORE WOOD. *Cheap Edition*, 2s. 6d.

Figuier's Popular Scientific Works. With Several Hundred Illustrations in each. 3s. 6d. each.

The Insect World.	Reptiles and Birds.	The Vegetable World.
The Human Race.	Mammalia.	The Ocean World.
	The World before the Deluge.	

Flora's Feast. A Masque of Flowers. Penned and Pictured by WALTER CRANE. With 40 pages in Colours. 5s.

Flower Painting, Elementary. With Eight Coloured Plates. 3s.

Flowers, and How to Paint Them. By MAUD NAFTEL. With Coloured Plates. 5s.

Football: the Rugby Union Game. Edited by Rev. F. MARSHALL. Illustrated. *New and Enlarged Edition.* 7s. 6d.

For Glory and Renown. By D. H. PARRY. Illustrated. *Cheap Edition.* 3s. 6d.

Fossil Reptiles, A History of British. By Sir RICHARD OWEN, F.R.S., &c. With 268 Plates. In Four Vols. £12 12s.

Franco-German War, Cassell's History of the. Complete in Two Vols., containing about 500 Illustrations. 9s. each.

Garden Flowers, Familiar. By F. E. HULME, F.L.S., F.S.A. With 200 Full-page Coloured Plates, and Descriptive Text by SHIRLEY HIBBERD. *Cheap Edition* In Five Vols., 3s. 6d. each.

Girl at Cobhurst, The. By FRANK R. STOCKTON. 6s.

Gladstone, The Right Hon. W. E., Cassell's Life of. Profusely Illustrated. 1s.
Gleanings from Popular Authors. With Original Illustrations. *Cheap Edition.* In One Vol., 3s. 6d.
Grace O'Malley, Princess and Pirate. By ROBERT MACHRAY. 6s.
Gulliver's Travels. With 88 Engravings. Cloth, 3s. 6d. ; cloth gilt, 5s.
Gun and its Development, The. By W. W. GREENER. With 500 Illustrations. *Entirely New Edition,* 10s. 6d.
Guns, Modern Shot. By W. W. GREENER. Illustrated. 5s.
Health, The Book of. By Eminent Physicians and Surgeons. Cloth, 21s.
Heavens, The Story of the. By Sir ROBERT STAWELL BALL, LL.D., F.R.S. With Coloured Plates and Wood Engravings. *Popular Edition,* 10s. 6d.
Heroes of Britain in Peace and War. With 300 Original Illustrations. *Cheap Edition.* Complete in One Vol., 3s. 6d.
History, A Foot-note to. Eight Years of Trouble in Samoa. By R. L. STEVENSON. 6s.
Home Life of the Ancient Greeks, The. Translated by ALICE ZIMMERN. Illustrated. *Cheap Edition.* 5s.
Horse, The Book of the. By SAMUEL SIDNEY. With 17 Full-page Collotype Plates of Celebrated Horses of the Day, and numerous other Illustrations. Cloth, 15s.
Horses and Dogs. By O. EERELMAN. With Descriptive Text. Translated from the Dutch by CLARA BELL. With Fifteen Full-page and other Illustrations. 25s. net.
Houghton, Lord : The Life, Letters, and Friendships of Richard Monckton Milnes, First Lord Houghton. By Sir WEMYSS REID. Two Vols. 32s.
Hygiene and Public Health. By B. ARTHUR WHITELEGGE, M.D. Illustrated. *New and Revised Edition.* 7s. 6d.
India, Cassell's History of. In One Vol. *Cheap Edition.* 7s. 6d.
In-door Amusements, Card Games, and Fireside Fun, Cassell's Book of. With numerous Illustrations. *Cheap Edition.* Cloth, 2s.
Iron Pirate, The. By MAX PEMBERTON. Illustrated. 5s.
Khiva, A Ride to. By Col. FRED BURNABY. *New Edition.* Illustrated. 3s. 6d.
King George, In the Days of. By Col. PERCY GROVES. Illustrated. 1s. 6d.
King Solomon's Mines. By H. RIDER HAGGARD. Illustrated. 3s. 6d. *People's Edition.* 6d.
Kronstadt. A New Novel. By MAX PEMBERTON. With 8 Full-page Plates. 6s.
Ladies' Physician, The. By a London Physician. *Cheap Edition, Revised and Enlarged.* 3s. 6d.
Lady's Dressing-Room, The. Translated from the French by Lady COLIN CAMPBELL. *Cheap Edition.* 2s. 6d.
Letts's Diaries and other Time-saving Publications are now published exclusively by CASSELL & COMPANY. (*A List sent post free on application.*)
Little Huguenot, The. *New Edition.* 1s. 6d.
Locomotive Engine, The Biography of a. By HENRY FRITH. 3s. 6d.
Loftus, Lord Augustus, P.C., G.C.B., The Diplomatic Reminiscences of. First Series. With Portrait. Two Vols. 32s. Second Series. Two Vols. 32s.
London, Cassell's Guide to. With Numerous Illustrations. 6d. Cloth, 1s.
London, Greater. By EDWARD WALFORD. Two Vols. With about 400 Illustrations. *Cheap Edition,* 4s. 6d. each. *Library Edition.* Two Vols. £1 the set.
London, Old and New. By WALTER THORNBURY and EDWARD WALFORD. Six Vols., with about 1,200 Illustrations. [*Cheap Ed.,* 4s. 6d. each. *Library Ed.,* £3.
Manchester, Old and New. By WILLIAM ARTHUR SHAW, M.A. With Original Illustrations. Three Vols., 31s. 6d.
Mechanics, The Practical Dictionary of. Three Vols., £3 3s. ; half-morocco, £3 15s. Supplementary Volume, £1 1s. ; or half morocco, £1 11s.
Medical Handbook of Life Assurance. By JAMES EDWARD POLLOCK, M.D., and JAMES CHISHOLM. *New and Revised Edition.* 7s. 6d.
Medicine, Manuals for Students of. (*A List forwarded post free on application.*)
Mesdag, H. W., the Painter of the North Sea. With Etchings and Descriptive Text. By PH. ZILCKEN. The Text translated from the Dutch by CLARA BELL. 36s.
Modern Europe, A History of. By C. A. FYFFE, M.A. *Cheap Edition in One Volume,* 10s. 6d. ; *Library Edition, Illustrated,* 3 vols., 7s. 6d. each.
Music, Illustrated History of. By EMIL NAUMANN. Edited by the Rev Sir F. A. GORE OUSELEY, Bart. Illustrated. Two Vols. 31s. 6d.

Selections from Cassell & Company's Publications.

National Library, Cassell's. Consisting of 214 Volumes. Paper covers, 3d. ; cloth, 6d. (*A Complete List of the Volumes post free on application.*)

Natural History, Cassell's Concise. By E. PERCEVAL WRIGHT, M.A., M.D., F.L.S. With several Hundred Illustrations. 7s. 6d.

Natural History, Cassell's New. Edited by P. MARTIN DUNCAN, M.B., F.R.S., F.G.S. *Cheap Edition.* With about 2,000 Illusts. Three Double Vols., 6s. each.

Nature and a Camera, With. Being the Adventures and Observations of a Field Naturalist and an Animal Photographer. By RICHARD KEARTON, F.Z.S. Illustrated by a Special Frontispiece, and 180 Pictures from Photographs by CHERRY KEARTON. 21s.

Nature's Wonder-Workers. By KATE R. LOVELL. Illustrated. 2s. 6d.

Nelson, The Life of. By ROBERT SOUTHEY. Illustrated with Eight Plates. 3s. 6d.

New Zealand, Pictorial. With Preface by Sir W. B. PERCEVAL, K.C.M.G. Illust. 6s.

Novels, Popular. Extra crown 8vo, cloth, 6s. each.

Grace O'Malley, Princess and Pirate. By ROBERT MACHRAY.
Cupid's Garden. By ELLEN THORNEYCROFT FOWLER.
A Limited Success. By SARAH PITT.
The Wrothams of Wrotham Court. By FRANCES HEATH FRESHFIELD.
Ill-gotten Gold : A Story of a Great Wrong and a Great Revenge. By W. G. TARBET.
Sentimental Tommy.
The Little Minister. } By J. M. BARRIE.
From the Memoirs of a Minister of France.
The Story of Francis Cludde. } By STANLEY WEYMAN.
Kronstadt.
Puritan's Wife, A.
The Impregnable City } By MAX PEMBERTON.
The Sea-Wolves.
Young Blood.
My Lord Duke.
Rogue's March, The. } By E. W. HORNUNG.
Spectre Gold.
By a Hair's-Breadth. } By HEADON HILL.
The Girl at Cobhurst.
Story Teller's Pack, A.
Mrs. Cliff's Yacht. } By FRANK STOCKTON.
The Adventures of Captain Horn.
Treasure Island. (Also People's Edition, 6d.)
The Master of Ballantrae.
The Black Arrow.
Kidnapped. } By ROBERT LOUIS STEVENSON. } Also a *Popular Edition.* 3s. 6d. each.
Catriona ; A Sequel to "Kidnapped."
Island Nights' Entertainments.
The Wrecker. By ROBERT LOUIS STEVENSON and LLOYD OSBOURNE.
What Cheer ! By W. CLARK RUSSELL.

Nursing for the Home and for the Hospital, A Handbook of. By CATHERINE J. WOOD. *Cheap Edition*, 1s. 6d. ; cloth, 2s.

Nursing of Sick Children, A Handbook for the. By CATHERINE J. WOOD. 2s. 6d.

Our Own Country. With 1,200 Illustrations. *Cheap Edition.* 3 Double Vols. 5s. each.

Painting, The English School of. By ERNEST CHESNEAU. *Cheap Edition*, 3s. 6d.

Paris, Old and New. Illustrated. In Two Vols. 9s. or 10s. 6d. each.

Peoples of the World, The. By Dr. ROBERT BROWN, F.L.S. Complete in Six Vols. With Illustrations. 7s. 6d. each.

Phrase and Fable, Dr. Brewer's Dictionary of. *Entirely New and largely increased Edition.* 10s. 6d. Also in half-morocco, 2 Vols., 15s.

Physiology for Students, Elementary. By ALFRED T. SCHOFIELD, M.D., M.R.C.S. With Two Coloured Plates and numerous Illustrations. *New Edition.* 5s.

Picturesque America. Complete in Four Vols., with 48 Exquisite Steel Plates, and about 800 Original Wood Engravings. £12 12s. the set. *Popular Edition* in Four Vols., price 18s. each.

Picturesque Australasia, Cassell's. With upwards of 1,000 Illustrations. In Four Vols., 7s. 6d. each.

Picturesque Canada. With about 600 Original Illustrations. 2 Vols. £9 9s. the set.

Picturesque Europe. Complete in Five Vols. Each containing 13 Exquisite Steel
. Plates, from Original Drawings, and nearly 200 Original Illustrations. £21. *Popular
Edition.* In Five Vols. 18s. each.

Picturesque Mediterranean, The. With a Series of Magnificent Illustrations
from Original Designs by leading Artists of the day. Two Vols. Cloth, £2 2s. each.

Pigeons, Fulton's Book of. Edited by LEWIS WRIGHT. Revised, Enlarged,
and Supplemented by the Rev. W. F. LUMLEY. With 50 Full-page Illustrations.
Popular Edition. In One Vol., 10s. 6d. *Original Edition*, with 50 Coloured Plates
and numerous Wood Engravings. 21s.

Planet, The Story of Our. By Prof. BONNEY, F.R.S., &c. With Coloured
Plates and Maps and about 100 Illustrations. *Cheap Edition.* 7s. 6d.

Polytechnic Series, The. Practical Illustrated Manuals. (*A List will be
sent on application.*)

Portrait Gallery, Cassell's Universal. Containing 240 Portraits of Celebrated
Men and Women of the Day. Cloth, 6s.

Portrait Gallery, The Cabinet. Complete in Five Series, each containing 36
Cabinet Photographs of Eminent Men and Women of the day. 15s. each.

Poultry, The Book of. By LEWIS WRIGHT. *Popular Edition.* Illustrated. 10s. 6d.

Poultry, The Illustrated Book of. By LEWIS WRIGHT. With Fifty Exquisite
Coloured Plates, and numerous Wood Engravings. *Revised Edition.* Cloth, gilt
edges, 21s.

"Punch," The History of. By M. H. SPIELMANN. With nearly 170 Illustrations, Portraits, and Facsimiles. Cloth, 16s. ; *Large Paper Edition*, £2 2s. net.

Q's Works, Uniform Edition of. 5s. each.

Dead Man's Rock.	The Astonishing History of Troy Town.
The Splendid Spur.	"I Saw Three Ships," and other Winter's Tales
The Blue Pavilions.	Noughts and Crosses.
The Delectable Duchy. Stories,	Wandering Heath.
Studies, and Sketches.	

Queen Summer ; or, The Tourney of the Lily and the Rose. Penned and
Portrayed by WALTER CRANE. With 40 pages in Colours. 6s.

Queen Victoria, The Life and Times of. By ROBERT WILSON. Complete in
2 Vols. With numerous Illustrations. 9s. each.

Queen's Empire, The. First Volume, containing about 300 Splendid Full-page
Illustrations. 9s.

Queen's London, The. Containing Exquisite Views of London and its Environs,
together with a fine series of Pictures of the Queen's Diamond Jubilee Procession.
Enlarged Edition. 10s. 6d.

Railway Guides, Official Illustrated. With Illustrations on nearly every page.
Maps &c. Paper covers, 1s.; cloth, 2s.

London and North Western Railway.	Great Eastern Railway.
Great Western Railway.	London and South Western Railway.
Midland Railway.	London, Brighton and South Coast Railway.
Great Northern Railway.	South Eastern Railway.

Abridged and Popular Editions of the above Guides can also be obtained. Paper covers. 3d. each

Railways, Our. Their Origin, Development, Incident, and Romance. By
JOHN PENDLETON. Illustrated. 2 Vols., 12s.

Rivers of Great Britain : Descriptive, Historical, Pictorial.
Rivers of the West Coast. With Etching as Frontispiece, and Numerous Illustrations in Text.
Royal 4to, 42s.
The Royal River: The Thames from Source to Sea. *Popular Edition*, 16s.
Rivers of the East Coast. With highly-finished Engravings. *Popular Edition* 16s.

Robinson Crusoe. *Cassell's New Fine-Art Edition.* With upwards of 100
Original Illustrations. *Cheap Edition*, 3s. 6d. or 5s.

Rogues of the Fiery Cross. By S. WALKEY. With 16 Full-page Illustrations. 5s

Ronner, Henriette, The Painter of Cat-Life and Cat-Character. By M. H.
SPIELMANN. Containing a Series of beautiful Phototype Illustrations. 12s.

Royal Academy Pictures. With upwards of 200 magnificent reproductions
of Pictures in the Royal Academy. 7s. 6d.

Russo-Turkish War, Cassell's History of. With about 400 Illustrations. *New
Edition.* In Two Vols., 9s. each.

Sala, George Augustus, The Life and Adventures of. By Himself. *Cheap
Edition.* One Vol., 7s. 6d.

Saturday Journal, Cassell's. Illustrated throughout. Yearly Vol., 7s. 6d.

Scarlet and Blue; or, Songs for Soldiers and Sailors. By JOHN FARMER. 5s. Words only, paper. 6d. ; cloth, 9d.

Science for All. Edited by Dr. ROBERT BROWN, M.A., F.L.S., &c. *Cheap Edition.* With over 1,700 Illustrations. Five Vols. 3s. 6d. each.

Science Series, The Century. Consisting of Biographies of Eminent Scientific Men of the present Century. Edited by Sir HENRY ROSCOE, D.C.L., F.R.S., M.P. Crown 8vo, 3s. 6d. each.

Michael Faraday, His Life and Work. By Professor SILVANUS P. THOMPSON, F.R.S
Pasteur. By PERCY FRANKLAND, F.R.S., and Mrs. FRANKLAND.
John Dalton and the Rise of Modern Chemistry. By Sir HENRY E; ROSCOE, F.R.S.
Major Ronnell, F.R.S. and the Rise of English Geography. By Sir CLEMENTS R. MARKHAM, C.B., F.R.S., President of the Royal Geographical Society.
Justus Von Liebig: His Life and Work. By W. A. SHENSTONE.
The Herschels and Modern Astronomy. By Miss AGNES M. CLERKE.
Charles Lyell and Modern Geology. By Professor T. G. BONNEY, F.R.S.
J. Clerk Maxwell and Modern Physics. By R. T. GLAZEBROOK, F.R.S.
Humphry Davy, Poet and Philosopher. By T. E. THORPE, F.R.S.
Charles Darwin and the Theory of Natural Selection. By EDWARD B. POULTON, M.A., F.R.S.

Scotland, Picturesque and Traditional. By G. E. EYRE-TODD. 6s.

Sea, The Story of the. An Entirely New and Original Work. Edited by Q. Illustrated. Complete in Two Vols., 9s. each. *Cheap Edition*, 5s. each.

Shaftesbury, The Seventh Earl of, K.G., The Life and Work of. By EDWIN HODDER. Illustrated. *Cheap Edition*, 3s. 6d.

Shakespeare, Cassell's Quarto Edition. Edited by CHARLES and MARY COWDEN CLARKE, and containing about 600 Illustrations by H. C. SELOUS. Complete in Three Vols., cloth gilt, £3 3s.—Also published in Three separate Vols., in cloth, viz. :—The COMEDIES, 21s. ; The HISTORICAL PLAYS, 18s. 6d. ; The TRAGEDIES, 25s.

Shakespeare, The England of. *New Edition.* By E. GOADBY. With Full-page Illustrations. Crown 8vo, 224 pages, 2s. 6d.

Shakespeare, The Plays of. Edited by Prof. HENRY MORLEY. Complete in 13 Vols., cloth, in box, 21s. ; also 39 Vols., cloth, in box, 21s. ; half-morocco, cloth sides, 42s.

Shakspere, The Leopold. With 400 Illustrations, and an Introduction by F. J. FURNIVALL. *Cheap Edition*, 3s. 6d. Cloth gilt, gilt edges, 5s. ; roxburgh, 7s. 6d.

Shakspere, The Royal. With Exquisite Steel Plates and Wood Engravings. Three Vols. 15s. each.

Sketches, The Art of Making and Using. From the French of G. FRAIPONT. By CLARA BELL. With Fifty Illustrations. 2s. 6d.

Social England. A Record of the Progress of the People. By various Writers. Edited by H. D. TRAILL, D.C.L. Complete in Six Vols. Vols. I. (Revised Ed.), II., and III., 15s. each. Vols. IV. and V., 17s. each. Vol. VI., 18s.

Spectre Gold. A Novel. By HEADON HILL. Illustrated. 6s.

Sports and Pastimes, Cassell's Complete Book of. *Cheap Edition*, 3s. 6d.

Star-Land. By Sir ROBERT STAWELL BALL, LL.D., &c. Illustrated. 6s.

Story of My Life, The. By the Rt. Hon. Sir RICHARD TEMPLE, Bart., G.C.S.I., &c. Two Vols. 21s.

Sun, The Story of the. By Sir ROBERT STAWELL BALL, LL.D., F.R.S., F.R.A.S. With Eight Coloured Plates and other Illustrations. *Cheap Edition*, 10s. 6d.

Taxation, Municipal, at Home and Abroad. By J. J. O'MEARA. 7s. 6d.

Thames, The Tidal. By GRANT ALLEN. With India Proof Impressions of 20 Magnificent Full-page Photogravure Plates, and many other Illustrations, after original drawings by W. L. WYLLIE, A.R.A. *New Edition*, cloth, 42s. net.

Things I have Seen and People I have Known. By G. A. SALA. With Portrait and Autograph. 2 Vols. 21s.

Three Homes, The. By the Very Rev. Dean FARRAR, D.D., F.R.S. *New Edition.* With 8 Full-page Illustrations. 6s

To the Death. By R. D. CHETWODE. With Four Plates. 5s.

Treatment, The Year-Book of, for 1898. A Critical Review for Practitioners of Medicine and Surgery. Fourteenth Year of Issue. 7s. 6d.

Trees, Familiar. By Prof. G. S. BOULGER, F.L.S., F.G.S. In Two Series. With Forty Coloured Plates in each. 12s. 6d. each.

Uncle Tom's Cabin. By HARRIET BEECHER STOWE. With upwards of 100 Original Illustrations. *Fine Art Memorial Edition.* 7s. 6d.

"Unicode": The Universal Telegraphic Phrase Book. Pocket or Desk Edition. 2s. 6d. each.

United States, Cassell's History of the. By EDMUND OLLIER. With 600 Illustrations. Three Vols. 9s. each.

Universal History, Cassell's Illustrated. With nearly ONE THOUSAND ILLUSTRATIONS. Vol. I. Early and Greek History.—Vol. II. The Roman Period.—Vol. III. The Middle Ages.—Vol. IV. Modern History. 9s. each.

Verses, Wise or Otherwise. By ELLEN THORNEYCROFT FOWLER. 3s. 6d.

War and Peace, Memories and Studies of. By ARCHIBALD FORBES, LL.D. *Original Edition,* 16s. *Cheap Edition,* 6s.

Water-Colour Painting, A Course of. With Twenty-four Coloured Plates by R. P. LEITCH, and full Instructions to the Pupil. 5s.

Westminster Abbey, Annals of. By E. T. BRADLEY (Mrs. A. MURRAY SMITH). Illustrated. With a Preface by the DEAN OF WESTMINSTER. 63s.

Wild Birds, Familiar. By W. SWAYSLAND. Four Series. With 40 Coloured Plates in each. (In sets only, price on application.)

Wild Flowers, Familiar. By F. E. HULME, F.L.S., F.S.A. With 200 Coloured Plates and Descriptive Text. *Cheap Edition.* In Five Vols., 3s. 6d. each.

Wild Flowers Collecting Book. In Six Parts, 4d. each.

Wild Flowers Drawing and Painting Book. In Six Parts. 4d. each.

Windsor Castle, The Governor's Guide to. By the Most Noble the MARQUIS OF LORNE, K.T. Profusely Illustrated. Limp cloth, 1s. Cloth boards, gilt edges. 2s.

World of Wit and Humour, Cassell's New. With New Pictures and New Text. Complete in Two Vols., 6s. each.

With Claymore and Bayonet. By Col. PERCY GROVES. With 8 Plates. 3s. 6d.

Work. The Illustrated Journal for Mechanics. Half-Yearly Vols. 4s. each.

"Work" Handbooks. A Series of Practical Manuals prepared under the Direction of PAUL N. HASLUCK, Editor of *Work.* Illustrated. Cloth, 1s. each.

World of Wonders, The. With 400 Illustrations. *Cheap Edition.* Two Vols., 4s. 6d. each.

Young Blood. A Novel. By E. W. HORNUNG. 6s.

ILLUSTRATED MAGAZINES.

The Quiver. Monthly, 6d.

Cassell's Magazine. Monthly, 6d.

"Little Folks" Magazine. Monthly, 6d.

The Magazine of Art. Monthly, 1s. 4d.

Cassell's Saturday Journal. Weekly, 1d.; Monthly, 6d.

Chums. The Illustrated Paper for Boys. Weekly, 1d.; Monthly, 6d.

Work. The Journal for Mechanics. Weekly, 1d.; Monthly, 6d.

Building World. Weekly, 1d.; Monthly, 6d.

Cottage Gardening. Illustrated. Weekly, ½d.; Monthly, 3d.

₄ *Full particulars of* CASSELL & COMPANY'S **Monthly Serial Publications** *will be found in* CASSELL & COMPANY'S COMPLETE CATALOGUE. •

Bibles and Religious Works.

Bible Biographies. Illustrated. 1s. 6d. each.
The Story of Joseph. Its Lessons for To-Day. By the Rev. GEORGE BAINTON.
The Story of Moses and Joshua. By the Rev. J. TELFORD.
The Story of Judges. By the Rev. J. WYCLIFFE GEDGE.
The Story of Samuel and Saul. By the Rev. D. C. TOVEY.
The Story of David. By the Rev. J. WILD.

The Story of Jesus. In Verse. By J. R. MACDUFF, D.D. 1s. 6d.

Bible, Cassell's Illustrated Family. With 900 Illustrations. Leather, gilt edges, £2 10s.; best full morocco, £3 15s.

Bible, Cassell's Guinea. With 900 Illustrations and Coloured Maps. Royal 4to. Leather, 21s. net. Persian antique, with corners and clasps, 25s. net.

Bible Educator, The. Edited by E. H. PLUMPTRE, D.D. With Illustrations, Maps, &c. Four Vols., cloth, 6s. each.

Bible Dictionary, Cassell's Concise. By the Rev. ROBERT HUNTER, LL.D., Illustrated. 7s. 6d.

Bible Student in the British Museum, The. By the Rev. J. G. KITCHIN, M.A. *Entirely New and Revised Edition,* 1s. 4d.

Bunyan, Cassell's Illustrated. With 200 Original Illustrations. *Cheap Edition,* 3s. 6d.

Bunyan's Pilgrim's Progress. Illustrated. *Cheap Edition,* cloth, 3s. 6d; cloth gilt, gilt edges, 5s.

Child's Bible, The. With 200 Illustrations. Demy 4to, 830 pp. *150th Thousand.* *Cheap Edition,* 7s. 6d. *Superior Edition,* with 6 Coloured Plates, gilt edges, 10s. 6d.

Child's Life of Christ, The. Complete in One Handsome Volume, with about 200 Original Illustrations. *Cheap Edition,* cloth, 7s. 6d.; or with 6 Coloured Plates, cloth, gilt edges, 10s. 6d.

Church of England, The. A History for the People. By the Very Rev. H. D. M. SPENCE, D.D., Dean of Gloucester. Illustrated. Vols. I., II., and III., 6s. each.

Church Reform in Spain and Portugal. By the Rev. H. E. NOYES, D.D. Illustrated. 2s. 6d.

Commentary for English Readers. Edited by Bishop ELLICOTT. With Contributions by eminent Scholars and Divines:—
New Testament. *Original Edition.* Three Vols., 21s. each; or in half morocco, £4 14s. 6d. the set. *Popular Edition.* Unabridged. Three Vols., 4s. each.
Old Testament. *Original Edition.* Five Vols., 21s. each; or in half-morocco, £7 17s. 6d. the set. *Popular Edition.* Unabridged. Five Vols., 4s. each.

° *The Complete Set of Eight Volumes in the Popular Edition is supplied at* 30s.

Commentary, The New Testament. Edited by Bishop ELLICOTT. Handy Volume Edition. Suitable for School and General Use.

St. Matthew. 3s. 6d.	**Romans.** 2s. 6d.	**Titus, Philemon, Hebrews,**
St. Mark. 3s.	**Corinthians I. and II.** 3s.	**and James.** 3s.
St. Luke. 3s. 6d.	**Galatians, Ephesians, and**	**Peter, Jude, and John.** 3s.
St. John. 3s. 6d.	**Philippians.** 3s.	**The Revelation.** 3s.
The Acts of the Apostles. 3s. 6d.	**Colossians, Thessalonians, and Timothy.** 3s.	**An Introduction to the New Testament.** 2s. 6d.

Commentary, The Old Testament. Edited by Bishop ELLICOTT. Handy Volume Edition. Suitable for School and General Use.

Genesis. 3s. 6d.	**Leviticus.** 3s.	**Deuteronomy.** 2s. 6d.
Exodus. 3s.	**Numbers.** 2s. 6d.	

Dictionary of Religion, The. An Encyclopædia of Christian and other Religious Doctrines, Denominations, Sects, Heresies, Ecclesiastical Terms, History, Biography, &c. &c. By the Rev. WILLIAM BENHAM, B.D. *Cheap Edition,* 10s. 6d.

Doré Bible. With 200 Full-page Illustrations by GUSTAVE DORÉ. *Popular Edition.* In One Vol. 15s. Also in leather binding. *(Price on application.)*

Early Days of Christianity, The. By the Very Rev. Dean FARRAR, D.D., F.R.S.
LIBRARY EDITION. Two Vols., 24s.; morocco. £2 2s.
POPULAR EDITION. In One Vol.; cloth, gilt edges, 7s. 6d.; tree-calf, 15s.
CHEAP EDITION. Cloth gilt, 3s. 6d.

Family Prayer-Book, The. Edited by the Rev. Canon GARBETT, M.A., and the Rev. S. MARTIN. With Full-page Illustrations. *New Edition.* Cloth. 7s. 6d.

"Graven in the Rock;" or, the Historical Accuracy of the Bible confirmed by reference to the Assyrian and Egyptian Sculptures in the British Museum and elsewhere. By the Rev. Dr. SAMUEL KINNS, F.R.A.S., &c. &c. Illustrated. *Library Edition*, in Two Volumes, cloth, with top edges gilded, 15s.

"Heart Chords." A Series of Works by Eminent Divines. In cloth, 1s. each.

My Father. By the Right Rev. Ashton Oxenden, late Bishop of Montreal.	**My Growth in Divine Life.** By the Rev. Prebendary Reynolds, M.A.
My Bible. By the Rt. Rev. W. Boyd Carpenter, Bishop of Ripon.	**My Hereafter.** By the Very Rev. Dean Bickersteth.
My Work for God. By the Right Rev. Bishop Cotterill.	**My Walk with God.** By the Very Rev. Dean Montgomery.
My Object in Life. By the Very Rev. Dean Farrar, D.D.	**My Aids to the Divine Life.** By the Very Rev. Dean Boyle.
My Aspirations. By the Rev. G. Matheson, D.D.	**My Sources of Strength.** By the Rev. E. E. Jenkins, M.A.
My Emotional Life. By Preb. Chadwick, D.D.	
My Body. By the Rev. Prof. W. G. Blaikie, D.D.	**My Comfort in Sorrow.** By Hugh Macmillan, D.D.
My Soul. By the Rev. P. B. Power, M.A.	

Helps to Belief. A Series of Helpful Manuals on the Religious Difficulties of the Day. Edited by the Rev. TEIGNMOUTH SHORE, M.A., Canon of Worcester, 1s. each.

CREATION. By Harvey Goodwin, D.D., late Lord Bishop of Carlisle.	PRAYER. By the Rev. Canon Shore, M.A.
MIRACLES. By the Rev. Brownlow Maitland, M.A.	THE ATONEMENT. By William Connor Magee, D.D., Late Archbishop of York.

Holy Land and the Bible, The. By the REV. CUNNINGHAM GEIKIE, D.D., LL.D. (Edin.). *Cheap Edition*, with 24 Collotype Plates, 12s. 6d.

Life of Christ, The. By the Very Rev. Dean FARRAR, D.D., F.R.S.
CHEAP EDITION. With 16 Full-page Plates. Cloth gilt, 3s. 6d.
POPULAR EDITION. With 16 Full-page Plates. Cloth gilt, gilt edges, 7s. 6d.
LARGE TYPE ILLUSTD. EDITION. Cloth, 7s. 6d. Cloth, full gilt, gilt edges, 10s. 6d.
LIBRARY EDITION. Two Vols. Cloth, 24s. morocco, 42s.

Methodism, Side-Lights on the Conflicts of, During the Second Quarter of the Nineteenth Century, 1827-1852. From the Notes of the Late Rev. JOSEPH FOWLER of the Debates of the Wesleyan Conference. Cloth. 8s.

Moses and Geology; or, the Harmony of the Bible with Science. By the Rev. SAMUEL KINNS, Ph.D., F.R.A.S. Illus. *Library Edition*, 10s. 6d.

New Light on the Bible and the Holy Land. By BASIL T. A. EVETTS, M.A. Illustrated. Cloth, 7s. 6d.

Old and New Testaments, Plain Introductions to the Books of the. Containing Contributions by many Eminent Divines. In Two Vols., 3s. 6d. each.

Plain Introductions to the Books of the Old Testament. 336 pages. Edited by Bishop ELLICOTT. 3s. 6d.

Plain Introductions to the Books of the New Testament. 304 pages. Edited by Bishop ELLICOTT. 3s. 6d.

Protestantism, The History of. By the Rev. J. A. WYLIE, LL.D. Containing upwards of 600 Original Illustrations. Three Vols., 27s.

"Quiver" Yearly Volume, The. With about 600 Original Illustrations and Coloured Frontispiece. 7s. 6d. Also Monthly, 6d.

St. George for England; and other Sermons preached to Children. *Fifth Edition.* By the Rev. T. TEIGNMOUTH SHORE, M.A., Canon of Worcester. 5s.

St. Paul, The Life and Work of. By the Very Rev. Dean FARRAR, D.D., F.R.S.
CHEAP ILLUSTRATED EDITION. 7s. 6d.
CHEAP EDITION. With 16 Full-page Plates, cloth gilt, 3s. 6d.
LIBRARY EDITION. Two Vols., cloth, 24s. ; calf, 42s.
ILLUSTRATED EDITION, One Vol., £1 1s. ; morocco, £2 2s.
POPULAR EDITION. Cloth, gilt edges, 7s. 6d.

Shortened Church Services and Hymns, suitable for use at Children's Services. Compiled by the Rev. Canon SHORE, *Enlarged Edition.* 1s.

"Six Hundred Years;" or, Historical Sketches of Eminent Men and Women who have more or less come into contact with the Abbey and Church of Holy Trinity, Minories, from 1293 to 1893, and some account of the Incumbents, the Fabric, the Plate, &c. &c. By the Vicar, the Rev. Dr. SAMUEL KINNS, F.R.A.S., &c. &c. With 65 Illustrations. 15s.

"Sunday:" Its Origin, History, and Present Obligation. By the Ven. Archdeacon HESSEY, D.C.L. *Fifth Edition*, 7s. 6d.

Educational Works and Students' Manuals.

Agricultural Text-Books, Cassell's. ('The " Downton " Series.) Fully Illustrated. Edited by JOHN WRIGHTSON, Professor of Agriculture. **Soils and Manures.** By J. M. H. MUNRO, D.Sc. (London), F.I.C., F.C.S. 2s. 6d. **Farm Crops.** By Professor WRIGHTSON. 2s. 6d. **Live Stock.** By Professor WRIGHTSON. 2s. 6d.

Alphabet, Cassell's Pictorial. Mounted on Linen, with Rollers. 2s. Mounted with Rollers, and Varnished. 2s. 6d.

Arithmetic :—Howard's Art of Reckoning. By C. F. HOWARD. Paper, 1s. ; cloth, 2s. *Enlarged Edition*, 5s.

Arithmetics, The "Belle Sauvage." By GEORGE RICKS, B.Sc. Lond. With Test Cards. (*List on application.*)

Atlas, Cassell's Popular. Containing 24 Coloured Maps. 1s. 6d.

Blackboard Drawing. By W. E. SPARKES. With 52 Full-page Illustrations by the Author. 5s.

Book-Keeping. By THEODORE JONES. FOR SCHOOLS, 2s. ; or cloth, 3s. FOR THE MILLION, 2s. ; or cloth, 3s. Books for Jones's System, Ruled Sets of, 2s.

British Empire Map of the World. New Map for Schools and Institutes. By G. R. PARKIN and J. G. BARTHOLOMEW, F.R.G.S. Mounted on cloth, varnished, and with Rollers or Folded. 25s.

Chemistry, The Public School. By J. H. ANDERSON, M.A. 2s. 6d.

Cookery for Schools. By LIZZIE HERITAGE. 6d.

Dulce Domum. Rhymes and Songs for Children. Edited by JOHN FARMER, Editor of "Gaudeamus," &c. Old Notation and Words, 5s. N.B.—The Words of the Songs in "Dulce Domum" (with the Airs both in Tonic Sol-Fa and Old Notation) can be had in Two Parts, 6d. each.

England, A History of. From the Landing of Julius Cæsar to the Present Day. By H. O. ARNOLD-FORSTER, M.P. Fully Illustrated. 5s.

English Literature, A First Sketch of, from the Earliest Period to the Present Time. By Prof. HENRY MORLEY. 7s. 6d.

Euclid, Cassell's. Edited by Prof. WALLACE, M.A. 1s.

Euclid, The First Four Books of. *New Edition.* In paper, 6d. ; cloth, 9d.

French, Cassell's Lessons in. *New and Revised Edition.* Parts I. and II., each 1s. 6d. ; complete, 2s. 6d. Key, 1s. 6d.

French-English and English-French Dictionary. 3s. 6d. or 5s.

French Reader, Cassell's Public School. By GUILLAUME S. CONRAD. 2s. 6d.

Galbraith and Haughton's Scientific Manuals.
   Astronomy. 5s. **Euclid.** Books I., II., III. 2s. 6d. Books IV., V., VI. 2s. 6d. **Mathematical Tables.** 3s. 6d. **Mechanics.** 3s. 6d. **Hydrostatics.** 3s. 6d. **Algebra.** Part I., cloth, 2s. 6d. Complete, 7s. 6d. **Tides and Tidal Currents,** with Tidal Cards, 3s.

Gaudeamus. Songs for Colleges and Schools. Edited by JOHN FARMER. 5s. Words only, paper, 6d. ; cloth, 9d.

Geometry, First Elements of Experimental. By PAUL BERT. Illustrated. 1s. 6d.

German Dictionary, Cassell's. German-English, English-German. *Cheap Edition*, cloth, 3s. 6d. ; half-morocco, 5s.

German Reading, First Lessons in. By A. JÄGST. Illustrated. 1s.

Hand and Eye Training. By GEORGE RICKS, B.Sc., and JOSEPH VAUGHAN. Illustrated. Vol. I. Designing with Coloured Papers. Vol. II. Cardboard Work, 2s. each. Vol. III. Colour Work and Design. 3s.

Hand and Eye Training. By G. RICKS, B.Sc. Two Vols., with 16 Coloured Plates in each. 6s. each. **Cards for Class Use.** Five Sets. 1s. each.

Historical Cartoons, Cassell's Coloured. Size 45 in. x 35 in., 2s. each. Mounted on canvas and varnished, with rollers, 5s. each. (Descriptive pamphlet, 16 pp., 1d.)

Italian Lessons, with Exercises, Cassell's. In One Vol. 2s.

Latin Dictionary, Cassell's. (Latin-English and English-Latin.) 3s. 6d. ; half morocco, 5s.

Latin Primer, The New. By Prof. J. P. POSTGATE. 2s. 6d.

Latin Primer, The First. By Prof. POSTGATE. 1s.

Latin Prose for Lower Forms. By M. A. BAYFIELD, M.A. 2s. 6d.

Laws of Every-Day Life. For the Use of Schools. By H. O. ARNOLD-FORSTER, M.P. 1s. 6d.

Lessons in Our Laws ; or, Talks at Broadacre Farm. By H. F. LESTER, B.A. In Two Parts. 1s. 6d. each.

Little Folks' History of England. By ISA CRAIG-KNOX. Illustrated. 1s. 6d.

Making of the Home, The. By Mrs. SAMUEL A. BARNETT. 1s. 6d.

Map Building for Schools. A Practical Method of Teaching Geography (England and Wales) By J. H. OVERTON, F.G.S. 6d.

Marlborough Books:—Arithmetic Examples. 3s. French Exercises. 3s. 6d. French Grammar. 2s. 6d. German Grammar. 3s. 6d.

Mechanics, Applied. By JOHN PERRY, M.E., D.Sc., &c. Illustrated. 7s. 6d.

Mechanics for Young Beginners. By the Rev. J. G. EASTON, M.A. *Cheap Edition,* 2s. 6d.

Mechanics and Machine Design, Numerical Examples in Practical. By R. G. BLAINE, M.E. *New Edition, Revised and Enlarged.* With 79 Illus. 2s. 6d.

Models and Common Objects, How to Draw from. By W. E. SPARKES. Illustrated. 3s.

Models, Common Objects, and Casts of Ornament, How to Shade from. By W. E. SPARKES. With 25 Plates by the Author. 3s.

Natural History Coloured Wall Sheets, Cassell's New. Consisting of 16 subjects. Size, 39 by 31 in. Mounted on rollers and varnished. 3s. each.

Object Lessons from Nature. By Prof. L. C. MIALL, F.L.S., F.G.S. Fully Illustrated. *New and Enlarged Edition.* Two Vols. 1s. 6d. each.

Physiology for Schools. By ALFRED T. SCHOFIELD, M.D., M.R.C.S., &c. Illustrated. 1s. 9d. Three Parts, paper covers, 5d. each ; or cloth limp, 6d. each.

Poetry Readers, Cassell's New. Illustrated. 12 Books. 1d. each. Cloth, 1s. 6d.

Popular Educator, Cassell's. With Revised Text, New Maps, New Coloured Plates, New Type, &c. Complete in Eight Vols., 5s. each.

Readers, Cassell's "Belle Sauvage." An Entirely New Series. Fully Illustrated. Strongly bound in cloth. *(List on application.)*

Reader, The Citizen. By H. O. ARNOLD-FORSTER, M.P. Cloth, 1s. 6d. ; also a Scottish Edition, cloth, 1s. 6d.

Readers, Cassell's Classical. Vol. I., 1s. 8d. ; Vol. II., 2s. 6d.

Reader, The Temperance. By J. DENNIS HIRD. 1s. or 1s. 6d.

Readers, Cassell's "Higher Class." *(List on application.)*

Readers, Cassell's Readable. Illustrated. *(List on application.)*

Readers for Infant Schools, Coloured. Three Books. 4d. each.

Readers, Geographical, Cassell's New. With Numerous Illustrations in each Book. *(List on application.)*

Readers, The Modern Geographical. Illustrated throughout. *(List on application.)*

Readers, The Modern School. Illustrated. *(List on application.)*

Rolit. An entirely novel system of learning French. By J. J. TYLOR. 3s.

Round the Empire. By G. R. PARKIN. With a Preface by the Rt. Hon. the Earl of Rosebery, K.G. Fully Illustrated. 1s. 6d.

Science of Every-Day Life. By J. A. BOWER. Illustrated. 1s.

Sculpture, A Primer of. By E. ROSCOE MULLINS. Illustrated. 2s. 6d.

Shakspere's Plays for School Use. Illustrated. 9 Books. 6d. each.

Spelling, A Complete Manual of. By J. D. MORELL, LL.D. 1s.

Technical Educator, Cassell's. A New Cyclopædia of Technical Education, with Coloured Plates and Engravings. Complete in Six Vols., 3s. 6d. each.

Technical Manuals, Cassell's. Illustrated throughout. 16 Vols., from 2s. to 4s. 6d. *(List free on application.)*

Technology, Manuals of. Edited by Prof. AYRTON, F.R.S., and RICHARD WORMELL, D.Sc., M.A. Illustrated throughout. *(List on application.)*

Things New and Old ; or, Stories from English History. By H. O. ARNOLD-FORSTER, M.P. Fully Illustrated. Strongly bound in cloth. Seven Books from 9d. to 1s. 8d.

World of Ours, This. By H. O. ARNOLD-FORSTER, M.P. Fully Illustrated. *Cheap Edition.* 2s. 6d.

Young Citizen, The ; or, Lessons in our Laws. By H. F. LESTER. Fully Illustrated. 2s. 6d.

Books for Young People.

Two Old Ladies, Two Foolish Fairies, and a Tom Cat. The Surprising Adventures of Tuppy and Tue. A New Fairy Story. By MAGGIE BROWNE. With Four Coloured Plates and Illustrations in text. Cloth, 3s. 6d.

Micky Magee's Menagerie; or, Strange Animals and their Doings. By S. H. HAMER. With 8 Coloured Plates and other Illustrations by HARRY NEILSON. Coloured Boards, 1s. 6d.

The Victoria Painting Book for Little Folks. Containing about 300 Illustrations suitable for Colouring, 1s.

"Little Folks" Half-Yearly Volume. Containing 480 pages of Letterpress, with Six Full-page Coloured Plates, and numerous other Pictures printed in Colour. Picture boards, 3s. 6d.; or cloth gilt, gilt edges, 5s.

Bo-Peep. A Treasury for the Little Ones. Yearly Vol. With Original Stories and Verses. Illustrated with Eight Full-page Coloured Plates, and numerous other Pictures printed in Colour. Elegant picture boards, 2s. 6d.; cloth, 3s. 6d.

Beneath the Banner. Being Narratives of Noble Lives and Brave Deeds. By F. J. CROSS. Illustrated. Limp cloth, 1s.; cloth boards, gilt edges, 2s.

Good Morning! Good Night! Morning and Evening Readings for Children, by the Author of "Beneath the Banner." Fully Illustrated. Limp cloth, 1s., or cloth boards, gilt edges, 2s.

Five Stars in a Little Pool. By EDITH CARRINGTON. Illustrated. 3s. 6d.

Merry Girls of England. By L. T. MEADE. 3s. 6d.

Beyond the Blue Mountains. By L. T. MEADE. Illustrated. 5s.

The Cost of a Mistake. By SARAH PITT. Illustrated. *New Edition.* 2s. 6d.

The Peep of Day. Cassell's Illustrated Edition. 2s. 6d.

A Book of Merry Tales. By MAGGIE BROWNE, SHEILA, ISABEL WILSON, and C. L. MATÉAUX. Illustrated. 3s. 6d.

A Sunday Story-Book. By MAGGIE BROWNE, SAM BROWNE, and AUNT ETHEL. Illustrated. 3s. 6d.

Story Poems for Young and Old. By E. DAVENPORT. 3s. 6d.

Pleasant Work for Busy Fingers. By MAGGIE BROWNE. Illustrated. 2s. 6d.

Magic at Home. By Prof. HOFFMAN. Fully Illustrated. A Series of easy and startling Conjuring Tricks for Beginners. Cloth gilt, 3s. 6d.

Little Mother Bunch. By Mrs. MOLESWORTH. Illustrated. *New Edition.* 2s. 6d.

Heroes of Every-Day Life. By LAURA LANE. With about 20 Full-page Illustrations. 256 pages, crown 8vo, cloth, 2s. 6d.

Ships, Sailors, and the Sea. By R. J. CORNEWALL-JONES. Illd. 2s. 6d.

Gift Books for Young People. By Popular Authors. With Four Original Illustrations in each. Cloth gilt, 1s. 6d. each.

The Boy Hunters of Kentucky. By Edward S. Ellis.	Jack Marston's Anchor.
Red Feather: a Tale of the American Frontier. By Edward S. Ellis.	Frank's Life-Battle.
Fritters; or, "It's a Long Lane that has no Turning."	Major Monk's Motto; or, "Look Before you Leap."
Trixy; or, "Those who Live in Glass Houses shouldn't throw Stones."	Tim Thomson's Trial; or, "All is not Gold that Glitters."
Rhoda's Reward.	Ursula's Stumbling-Block.
	Ruth's Life-Work; or,"No Pains, no Gains."
	Uncle William's Charge.

"Golden Mottoes" Series, The. Each Book containing 208 pages, with Four Full-page Original Illustrations. Crown 8vo, cloth gilt, 2s. each.

"Nil Desperandum." By the Rev. F. Langbridge, M.A.	"Honour is my Guide." By Jeanie Hering (Mrs. Adams-Acton).
"Bear and Forbear." By Sarah Pitt.	"Aim at a Sure End." By Emily Searchfield.
"Foremost if I Can." By Helen Atteridge.	"He Conquers who Endures." By the Author of "May Cunningham's Trial." &c.

"Cross and Crown" Series, The. With Four Illustrations in each Book. Crown 8vo, 256 pages, 2s. 6d. each.

Heroes of the Indian Empire; or, Stories of Valour and Victory. By Ernest Foster.	Adam Hepburn's Vow: A Tale of Kirk and Covenant. By Annie S. Swan.
Through Trial to Triumph; or, "The Royal Way." By Madeline Bonavia Hunt.	No. XIII.; or, The Story of the Lost Vestal. A Tale of Early Christian Days. By Emma Marshall.
Strong to Suffer: A Story of the Jews. By E. Wynne.	
By Fire and Sword: A Story of the Huguenots. By Thomas Archer.	Freedom's Sword: A Story of the Days of Wallace and Bruce. By Annie S. Swan.